Lecture Notes in Computer Science 15699

Founding Editors

Gerhard Goos
Juris Hartmanis

Editorial Board Members

Elisa Bertino, *Purdue University, West Lafayette, IN, USA*
Wen Gao, *Peking University, Beijing, China*
Bernhard Steffen, *TU Dortmund University, Dortmund, Germany*
Moti Yung, *Columbia University, New York, NY, USA*

The series Lecture Notes in Computer Science (LNCS), including its subseries Lecture Notes in Artificial Intelligence (LNAI) and Lecture Notes in Bioinformatics (LNBI), has established itself as a medium for the publication of new developments in computer science and information technology research, teaching, and education.

LNCS enjoys close cooperation with the computer science R & D community, the series counts many renowned academics among its volume editors and paper authors, and collaborates with prestigious societies. Its mission is to serve this international community by providing an invaluable service, mainly focused on the publication of conference and workshop proceedings and postproceedings. LNCS commenced publication in 1973.

Mindaugas Bloznelis · Paulius Drungilas ·
Bogumił Kamiński · Paweł Prałat ·
Matas Šileikis · François Théberge ·
Rimantas Vaicekauskas
Editors

Modelling and Mining Networks

20th International Workshop, WAW 2025
Vilnius, Lithuania, June 30 – July 3, 2025
Proceedings

Editors
Mindaugas Bloznelis
Vilnius University
Vilnius, Lithuania

Bogumił Kamiński
SGH Warsaw School of Economics
Warsaw, Poland

Matas Šileikis
Czech Academy of Sciences
Prague, Czech Republic

Rimantas Vaicekauskas
Vilnius University
Vilnius, Lithuania

Paulius Drungilas
Vilnius University
Vilnius, Lithuania

Paweł Prałat
Toronto Metropolitan University
Toronto, ON, Canada

François Théberge
Tutte Institute for Mathematics
and Computing
Ottawa, ON, Canada

ISSN 0302-9743　　　　　　　　ISSN 1611-3349　(electronic)
Lecture Notes in Computer Science
ISBN 978-3-031-92897-0　　　　ISBN 978-3-031-92898-7　(eBook)
https://doi.org/10.1007/978-3-031-92898-7

© The Editor(s) (if applicable) and The Author(s), under exclusive license
to Springer Nature Switzerland AG 2025

This work is subject to copyright. All rights are solely and exclusively licensed by the Publisher, whether the whole or part of the material is concerned, specifically the rights of translation, reprinting, reuse of illustrations, recitation, broadcasting, reproduction on microfilms or in any other physical way, and transmission or information storage and retrieval, electronic adaptation, computer software, or by similar or dissimilar methodology now known or hereafter developed.
The use of general descriptive names, registered names, trademarks, service marks, etc. in this publication does not imply, even in the absence of a specific statement, that such names are exempt from the relevant protective laws and regulations and therefore free for general use.
The publisher, the authors and the editors are safe to assume that the advice and information in this book are believed to be true and accurate at the date of publication. Neither the publisher nor the authors or the editors give a warranty, expressed or implied, with respect to the material contained herein or for any errors or omissions that may have been made. The publisher remains neutral with regard to jurisdictional claims in published maps and institutional affiliations.

This Springer imprint is published by the registered company Springer Nature Switzerland AG
The registered company address is: Gewerbestrasse 11, 6330 Cham, Switzerland

If disposing of this product, please recycle the paper.

Preface

The *20th Workshop on Modelling and Mining Networks (WAW 2025)* was held at the Vilnius University, Vilnius, Lithuania (June 30 – July 3, 2025). This is an annual meeting, providing opportunities for researchers in network science to interact and to exchange research ideas. We do hope that the event was an effective venue for the dissemination of new results and for fostering research collaboration.

Virtually every human-technology interaction, or sensor network, generates observations that are in some relation with each other. As a result, many data science problems can be viewed as a study of some properties of complex networks in which nodes represent the entities that are being studied and edges represent relations between these entities. Such networks are often large-scale, decentralized, and evolve dynamically over time. Modelling and mining complex networks in order to understand the principles governing the organization and the behaviour of such networks is crucial for a broad range of fields of study, including information and social sciences, economics, biology, and neuroscience.

The aim of the *20th Workshop on Modelling and Mining Networks (WAW 2025)* was to further the understanding of networks that arise in theoretical as well as applied domains. The goal was also to stimulate the development of high-performance and scalable algorithms that exploit these networks. The workshop welcomes the researchers who are working on graph-theoretic and algorithmic aspects of networks represented as graphs or hypergraphs and other higher order structures.

This volume contains the papers accepted and presented during the workshop. Each submission was carefully reviewed by the members of the Programme Committee. Papers were submitted and reviewed using the EasyChair online system. The committee members decided to accept 13 papers.

July 2025

Mindaugas Bloznelis
Paulius Drungilas
Bogumił Kamiński
Paweł Prałat
Matas Šileikis
François Théberge
Rimantas Vaicekauskas

Organization

General Chairs

Andrei Z. Broder — Google Research, USA
Fan Chung Graham — University of California San Diego, USA
Paweł Prałat — Toronto Metropolitan University, Canada

Organizing Committee

Mindaugas Bloznelis — Vilnius University, Lithuania
Paulius Drungilas — Vilnius University, Lithuania
Bogumił Kamiński — SGH Warsaw School of Economics, Poland
Paweł Prałat — Toronto Metropolitan University, Canada
Matas Šileikis — Czech Academy of Sciences, Czech Republic
François Théberge — Tutte Institute for Mathematics and Computing, Canada
Rimantas Vaicekauskas — Vilnius University, Lithuania

Sponsoring Institutions

Vilnius University, Lithuania
Toronto Metropolitan University, Canada
Tutte Institute for Mathematics and Computing, Canada
Lithuanian Mathematical Society, Lithuania
Go Vilnius, Lithuania
Euromonitor International, UK
Google, USA

Program Committee

Konstantin Avratchenkov — Inria, France
Leman Akoglu — Carnegie Mellon University, USA
Mindaugas Bloznelis — Vilnius University, Lithuania
Paolo Boldi — University of Milan, Italy
Anthony Bonato — Toronto Metropolitan University, Canada

Ulrik Brandes	ETH Zürich, Switzerland
Piotr Bródka	Wrocław University of Science and Technology, Poland
Hocine Cherifi	University of Burgundy, France
Fan Chung Graham	UC San Diego, USA
Collin Cooper	King's College London, UK
Megan Dewar	Tutte Institute for Mathematics and Computing, Canada
Andrzej Dudek	Western Michigan University, USA
Alan Frieze	Carnegie Mellon University, USA
Jeannette Janssen	Dalhousie University, Canada
Cliff Joslyn	Pacific Northwest National Laboratory, USA
Bogumił Kamiński	SGH Warsaw School of Economics, Poland
Julia Komjathy	Delft University of Technology, The Netherlands
Ravi Kumar	Google, USA
Nicolas Landry	University of Vermont, USA
Marc Lelarge	Inria, France
Lasse Leskela	Aalto University, Finland
Nelly Litvak	University of Twente, The Netherlands
Oliver Mason	NUI Maynooth, Ireland
Paweł Misiorek	Poznań University of Technology, Poland
Dieter Mitsche	Universitat Politècnica de Catalunya, Spain
Peter Morters	University of Cologne, Germany
Tobias Mueller	Groningen University, The Netherlands
Mariana Olvera-Cravioto	University of North Carolina at Chapel Hill, USA
Pan Peng	University of Science and Technology of China, China
Xavier Perez-Gimenez	University of Nebraska-Lincoln, USA
Paweł Prałat	Toronto Metropolitan University, Canada
Katarzyna Rybarczyk	Adam Mickiewicz University, Poland
Clara Stegehuis	University of Twente, The Netherlands
Przemysław Szufel	SGH Warsaw School of Economics, Poland
François Théberge	Tutte Institute for Mathematics and Computing, Canada
Remco van der Hofstad	Eindhoven University of Technology, The Netherlands
Yana Volkovich	Microsoft, USA
Stephen Young	Pacific Northwest National Laboratory, USA
Yue Zhao	University of Southern California, USA

Contents

Linear Geometric Centralities 1
   Paolo Boldi, Flavio Furia, and Chiara Prezioso

Analysis and Predictability of Centrality Measures in Competition
Networks 17
   Anthony Bonato and Mariam Walaa

Modularity of Random Intersection Graphs 30
   Katarzyna Rybarczyk

The Size of the Giant in Inhomogeneous Random Graphs of Preferential
Attachment Type 45
   Peter Mörters and Lucas Schätze

k-Connectivity Threshold for Superpositions of Bernoulli Random Graphs 65
   Daumilas Ardickas, Mindaugas Bloznelis, and Rimantas Vaicekauskas

Improving Community Detection via Community Association Strength
Scores 81
   *Jordan Barrett, Ryan DeWolfe, Bogumił Kamiński, Paweł Prałat,
   Aaron Smith, and François Théberge*

PageRank Under Interpolation Between Undirected- and Directed
Networks - A Case Study 96
   Florian Henning, Remco van der Hofstad, and Nelly Litvak

Degrees in Preferential Attachment Networks with an Anomaly 109
   Qiu Liang, Remco van der Hofstad, and Nelly Litvak

The Artificial Benchmark for Community Detection with Outliers
and Overlapping Communities (**ABCD**+o^2) 125
   *Jordan Barrett, Ryan DeWolfe, Bogumił Kamiński, Paweł Prałat,
   Aaron Smith, and François Théberge*

A Graph Network Approach to Disinformation Detection in Social Media 141
   *Milita Songailaitė, Justina Mandravickaitė, Veronika Bryskina,
   Maksym Bondar, and Tomas Krilavičius*

Computation of the Laplacian Spectral Barycentre Network in a Soules
Basis 157
  François G. Meyer

The Multilayer Artificial Benchmark for Community Detection (**mABCD**) 172
  Piotr Bródka, Michał Czuba, Bogumił Kamiński, Łukasz Kraiński,
  Paweł Prałat, and François Théberge

Integrating Link Prediction and Isolation Forest for Backbone Extraction 189
  Ali Yassin, Hocine Cherifi, Hamida Seba, and Olivier Togni

Author Index 205

Linear Geometric Centralities

Paolo Boldi, Flavio Furia, and Chiara Prezioso

Dipartimento di Informatica, Università degli Studi di Milano, Milan, Italy
paolo.boldi@unimi.it

Abstract. Centrality indices are ways to measure the importance of nodes in a graph; this need is so obviously relevant that it was discussed many times in sociology, psychology, mathematics and computer science, giving rise to a whole zoo of definitions of centrality. The ideas underlying such definitions are wildly different, but many centrality measures are based on shortest-path distances: such centralities are referred to as *geometric*. Albeit geometric centralities can use the shortest-path-length information in many different ways, most of the existing geometric centralities can be defined as a linear transformation. In this paper we define formally the class of linear geometric centralities in its full generality, and study its main properties and expressivity.

Keywords: network centrality · linear transformation · shortest paths · Farkas' lemma

1 Introduction and Motivation

Network analysis is a fascinating field, as studying human-created networks offers quantitative insights into properties that influence the domain that the network represents. One interesting issue is to determine which nodes in a graph are the most important: centrality indices are used to assess quantitatively how "central" each node is in a graph, under many possible interpretations of the meaning of importance.

In this work, we study indices that can be referred to as *geometric*, as they solely depend on how many nodes exist at every (shortest-path) distance. Notable examples in this category include in-degree, closeness [3,4], and harmonic [6] centrality. In this paper, in particular, we propose and study a class of geometric measures that can be computed as linear combination of the number of nodes at every distance: these centralities will be referred to as *linear geometric centralities* (or just linear centralities, for short).

A special class of linear centralities are those assigning larger weights to shorter distances, based on the intuition that a node with many nearby connections is more important than a node with many distant ones. These types of decaying centralities first appeared in [12] (albeit with the equivalent but opposite postulation that smaller centrality values denote a higher importance). In [13] the same class was extended to directed graphs, and several of its properties were explored.

In more recent years, decaying centralities appeared in [18], where they are called additive and studied through the so-called "axiomatic approach" (pioneered in [15]). Finally, in [5] the axiomatic approach was used to investigate the impact of edge additions to the final scores and rankings they yield.

In the present paper, we relax the constraint of decaying weights, and study linear geometric centralities in their full generality: our aim is to analyze their expressivity and to understand whether they capture a broader range of behaviors than decaying centralities.

In particular, the main goal of this work is to answer the two following questions: (a) given a pair of linear geometric centralities, is there always a graph on which they rank nodes differently? (b) given a graph, what is the maximum number of rankings of its nodes that linear centralities can induce? how can we characterize precisely those rankings?

To the best of our knowledge, this is the first attempt at studying this wide class of centrality measures. This work, focusing especially on expressiveness, follows the path opened, for instance, by [16] in that it attempts to find general results for large sets of centralities.

2 Notations and Basic Definitions

For every natural number n, we let $[n] = \{0, 1, \ldots, n-1\}$.

Graphs. A *graph* $G = (V_G, E_G)$ is given [7] by a finite set of nodes V_G and[1] a set of arcs $E_G \subseteq V_G \times V_G$. We shall always assume, without loss of generality, that $V = [n]$ where n is the number of nodes.

We write $x \to y$ to mean that $x, y \in V$ and $(x, y) \in E$. A graph is *undirected* iff $x \to y$ implies $y \to x$. When drawing undirected graphs, pairs of opposite arcs are represented as a single undirected edge.

A *path* of length k from $x \in V$ to $y \in V$ is a sequence $(x_0, x_1, \ldots, x_k)$ of nodes such that $x = x_0$, $y = x_k$ and $x_i \to x_{i+1}$ for all $i \in [k]$. The *(shortest path) distance* from x to y in G, denoted by $d_G(x, y)$, is the length of a shortest path from x to y, or ∞ if no path from x to y exists.

Note that in this paper, unless otherwise specified, we will consider *directed* graphs, where $d(x, y)$ and $d(y, x)$ can be different (it can even be the case that one is finite and the other is not!).

Graph Homomorphisms. A *graph homomorphism* $\varphi : G \to H$ is a function $\varphi : V_G \to V_H$ such that $x \to_G y$ iff $\varphi(x) \to_H \varphi(y)$, for all $x, y \in V_G$. A bijective homomorphism $\varphi : G \to H$ is called an *isomorphism*; it is an *automorphism* if $G = H$. A graph with no non-trivial automorphism (i.e., whose only automorphism is the identity) is called *rigid*.

---

[1] For this and similar notations, the subscript G is dropped whenever it is clear from the context.

Matrices and Vectors. In this paper, we shall use column vectors; all vectors and matrices are conveniently indexed starting from 0. So, if $\mathbf{x} \in \mathbf{R}^n$ and $A \in \mathbf{R}^{m \times n}$ then

$$\mathbf{x} = \begin{bmatrix} x_0 \\ x_1 \\ \vdots \\ x_{n-1} \end{bmatrix} \quad \text{and} \quad A = \begin{bmatrix} a_{0,0} & a_{0,1} & \cdots & a_{0,n-1} \\ a_{1,0} & a_{1,1} & \cdots & a_{1,n-1} \\ \vdots & \vdots & \ddots & \vdots \\ a_{m-1,0} & a_{m-1,1} & \cdots & a_{m-1,n-1} \end{bmatrix}$$

Infinite vectors and matrix product. If $\mathbf{a} \in \mathbf{R}^{\mathbf{N}}$, we view $\mathbf{a}$ as an infinite vector of real numbers. If $A \in \mathbf{R}^{m \times n}$, we write $A \cdot \mathbf{a}$ as a shortcut for

$$A \cdot \mathbf{a}[:n]$$

where $\mathbf{a}[:n]$ stands for the vector of the first n entries of $\mathbf{a}$.

Permutations. For every n, we let S_n be the symmetric group of all permutations (i.e., bijections) $\pi : [n] \to [n]$. The permutation matrix R_π is the binary $n \times n$ matrix such that

$$(R_\pi)_{ij} = 1 \text{ iff } \pi(i) = j.$$

3 Centralities, Permutations and Agreements

A *(graph) centrality* $\mathfrak{f}$ [8] is a function associating with each graph G a map $\mathfrak{f}_G : V_G \to \mathbf{R}$ such that for any two graphs G, H, if $\varphi : G \to H$ is an isomorphism then

$$\mathfrak{f}_G(x) = \mathfrak{f}_H(\varphi(x))$$

for all $x \in V_G$.

While usually centralities are restricted to having non-negative values only, here we are going to relax this requirement and allow for negative centrality values. As usual, the value $\mathfrak{f}_G(x)$ is interpreted as a score of how central (i.e., important) node x is in G: larger values imply greater importance.

We say that $\mathfrak{f}$ has *no ties on* G iff $\mathfrak{f}_G(x) \neq \mathfrak{f}_G(y)$ whenever $x, y \in V_G$ and $x \neq y$. Note that a centrality always has ties on non-rigid graphs: if $\phi : G \to G$ is an automorphism then by definition $\mathfrak{f}(\phi(x)) = \mathfrak{f}(x)$ for all x such that $\phi(x) \neq x$.

Here are some notable examples of centralities[2] (see, for instance, [6] for further examples and taxonomy):

- in-degree: $\mathfrak{C}_G^{\text{in}}(x) = |\{y \mid y \to x\}|$;
- closeness [3]: $\mathfrak{C}_G^{\text{cl}}(x) = 1/\sum_{y \in V, d(y,x) < \infty} d(y,x)$;
- Lin [14]: $\mathfrak{C}_G^{\text{Lin}}(x) = |\{y \in V \mid d(y,x) < \infty\}|^2 / \sum_{y \in V, d(y,x) < \infty} d(y,x)$;

---

[2] Some of these definitions were originally given for undirected graphs only. Here we are providing the adaptations for the general (i.e., directed) case, with the usual proviso that incoming paths are more interesting than outgoing paths.

- harmonic [6]: $\mathfrak{C}_G^{\mathrm{harm}}(x) = \sum_{y \in V} 1/d(y,x)$, with the proviso that $1/\infty = 0$;
- Katz [11]: $\mathfrak{C}_G^{\mathrm{Katz}}(x) = \sum_\pi \beta^{|\pi|}$, where the sum ranges over all paths π ending in x, and β is a parameter smaller than the reciprocal of the spectral radius of G;
- betweenness [1]: $\mathfrak{C}_G^{\mathrm{bet}}(x) = \sum_{y,z,\sigma_{yz} \neq 0} \sigma_{yz}(x)/\sigma_{yz}$, where the sum ranges over all pairs of nodes $y \neq z$, σ_{yz} is the number of shortest paths from y to z, and $\sigma_{yz}(x)$ is the number of such paths passing through x.

In most cases, more than the actual value a centrality assigns to each node, we are interested in how nodes are ranked with respect to a given centrality. This concept is captured formally by the following:

Definition 1 (Respecting a permutation). *Given a graph G of n nodes, we say that $\mathfrak{f}$ respects the permutation $\pi \in S_n$ on G iff $\mathfrak{f}_G(\pi(i)) \geq \mathfrak{f}_G(\pi(i+1))$ for all $i \in [n-1]$ (that is, π orders the nodes of G in non-increasing order of their centrality values with respect to $\mathfrak{f}$).*

Note that if $\mathfrak{f}$ has no ties on G then there is a unique permutation that $\mathfrak{f}$ respects on G, called the *permutation induced by* $\mathfrak{f}$.

Definition 2 (Agreement between centralities). *Two centralities $\mathfrak{f}$ and $\mathfrak{g}$ agree on a graph G of n nodes iff they respect the same permutations on G. (In particular, if they both have no ties on G, then they must induce the same permutation). Two centralities that agree on all graphs are called* equivalent.

For example, the following two centralities:

$$\mathfrak{C}_G^{\mathrm{ccl}}(x) = \frac{|V|}{\sum_{y \in V, d(y,x) < \infty} d(y,x)} \qquad \text{"classical" closeness}$$

$$\mathfrak{C}_G^{\mathrm{np}}(x) = - \sum_{y \in V, d(y,x) < \infty} d(y,x) \qquad \text{negative peripherality}$$

are both equivalent to $\mathfrak{C}_G^{\mathrm{cl}}$ (closeness). Classical closeness differs from closeness only by a positive multiplicative constant $|V|$, that has no impact on the ranking of nodes. For negative peripherality, instead of taking the reciprocal of the sum of distances to x, we take the *opposite* of the sum of distances: although the actual values are no doubt different, $\mathfrak{C}_G^{\mathrm{np}}$ induces the same orders on the nodes as the one induced by $\mathfrak{C}_G^{\mathrm{cl}}$.

4 Geometric and Linear Centralities

In the wide world of centralities, many depend only on the number of nodes at a(ny) given distance. As explained in Sect. 1, in this paper we focus on this class.

Definition 3 (Distance-count function). *Given a graph G with n nodes, and a node $i \in V$, let the* distance-count function *of G at i be the function $c_{G,i} : \mathbf{N} \to \mathbf{N}$ defined by*

$$c_{G,i}(k) = |\{j \in V : d(j,i) = k\}|,$$

that is, the number of nodes at distance k to i.

Note that we do not take infinite distances into account (differently from [13]), thus $c_{G,i}(k) = 0$ for all $k \geq n$, because two nodes cannot have *finite* distance larger than $n - 1$. Moreover, note also that $c_{G,i}(0) = 1$ always, because there is exactly one node at distance 0 from i, that is, i itself.

A centrality is geometric if it is essentially dependent only on distance counts:

Definition 4 (Geometric centralities). *A centrality* $\mathfrak{f}$ *is strictly geometric iff for all graphs G and G' and all nodes $i \in V_G$ and $i' \in V_{G'}$ we have*[3]

$$c_{G,i} \equiv c_{G',i'} \text{ implies } \mathfrak{f}_G(i) = \mathfrak{f}_{G'}(i').$$

A centrality $\mathfrak{f}$ *is geometric iff it is equivalent to one that is strictly geometric.*

For instance, in-degree, closeness, Lin, negative peripherality and harmonic are all strictly geometric: their value on a node x only depends on how many nodes are at each distance from x. Classical closeness ($\mathfrak{C}_G^{\mathrm{ccl}}$) is not strictly geometric: to compute its value on x you need to know the number of nodes in the graph, which is not deducible only from the distance-count function unless the graph is strongly connected. Nonetheless, $\mathfrak{C}_G^{\mathrm{ccl}}$ is equivalent to closeness ($\mathfrak{C}_G^{\mathrm{cl}}$), so $\mathfrak{C}_G^{\mathrm{ccl}}$ is geometric.

Conversely, Katz centrality is not geometric (because it does not depend on shortest paths, but on *all* paths), neither is betweenness (because it does not depend on the length of shortest paths, but on their number).

Geometric centralities depend only on distance counts, but they may do so in many different ways. We are now going to focus further our attention on the case where the dependence is linear. Let us first define:

Definition 5 (Distance-count matrix). *The* distance-count matrix *of a graph G of n nodes is the matrix $C = C_G \in \mathbf{R}^{n \times n}$ such that the i-th row of C is $\mathbf{c}_{G,i}[:n]$. Said otherwise, $c_{i,k}$ is the number of nodes at distance k to i, for all $i, k \in [n]$.*

In Fig. 1, we show an example of distance-count matrix: node 0 (first row) has 1 node at distance 0 (itself), and $n - 1$ nodes at distance 1; node 1 (second row) has 1 node at distance 0, 1 node at distance 1 (node 0), and $n - 2$ nodes at distance 2; and so on.

Definition 6 (Linear centrality). *Given* $\mathbf{a} \in \mathbf{R}^N$ *(the* coefficient vector*), the centrality $\mathfrak{L}_G^{\mathbf{a}}$ is defined by*

$$\mathfrak{L}_G^{\mathbf{a}}(i) = (C_G \cdot \mathbf{a})_i.$$

A centrality is strictly linear (geometric) *if it is $\mathfrak{L}_G^{\mathbf{a}}$ for some $\mathbf{a} \in \mathbf{R}^N$. A centrality is* linear (geometric) *if it is equivalent to a strictly linear (geometric) centrality.*

---
[3] We use $\equiv$ to denote that the two functions are equal.

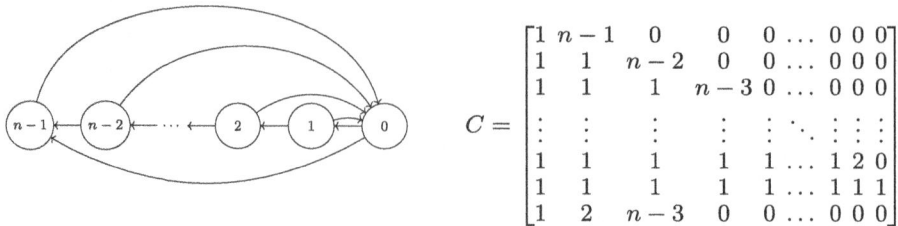

Fig. 1. A graph with n nodes, and its distance-count matrix C.

For instance,

- in-degree is strictly linear, with coefficients $(0, 1, 0, 0, \dots)^T$: its value on x is exactly the number of in-neighbors of x (i.e., nodes at distance 1 to x);
- negative peripherality is strictly linear, with coefficients $(0, -1, -2, -3, \dots)^T$;
- harmonic is strictly linear, with coefficients $(0, 1, 1/2, 1/3, 1/4, \dots)^T$;
- closeness and classic closeness are not strictly linear, but they are all linear, since they are all equivalent to negative peripherality.

5 Centralities, Ties and Rigidity

By the very definition of centrality, every centrality (including of course all geometric centralities) has ties on all non-rigid graphs. We can say something more:

Theorem 1. *All centrality measures have ties on graphs that are not rigid. On the other hand, there exists a centrality measure $\mathfrak{f}$ such that, for all rigid graphs G, $\mathfrak{f}$ has no ties on G.*

Proof. We exploit the lexicographic canonization process described in [2]: let G be a rigid graph with n nodes, and A_π be the adjacency matrix of G when its nodes are permuted according to $\pi \in S_n$. Suppose that $A_\pi = A_\rho$ for some choice of permutations $\pi, \rho \in S_n$; then for all $x, y \in V_G$:

$$A_{\pi(x)\pi(y)} = A_{\rho(x)\rho(y)},$$

that is, $\pi(x) \to_G \pi(y)$ iff $\rho(x) \to_G \rho(y)$. But then $\rho^{-1} \circ \pi$ would be an automorphism of G, hence $\pi = \rho$. So, the matrices A_π are all distinct; hence, there is a unique π such that A_π is minimal in lexicographic order (we can order matrices lexicographically by considering the string obtained concatenating all of their rows). Now you can define $\mathfrak{f}_G$ to be such $\pi : V_G \to [n]$: since π is a permutation, by definition it is injective. □

Geometric centralities, though, are more limited and may give rise to unsolvable ties even on rigid graphs. For instance, the graph in Fig. 2 has no non-trivial automorphisms, so according to Theorem 1 there is a centrality that produces no ties on it; nonetheless, all geometric centralities give the same value to nodes 1 and 2, because the corresponding rows in its distance-count matrix C_G are the same. Let us provide a formal definition of this limitation:

Definition 7 (Geometrically rigid). *A graph G is geometrically rigid if the rows of its distance-count matrix C_G are all distinct.*

Clearly, every geometrically rigid graph is also rigid, but not the other way round, as the graph in Fig. 2 shows. Nonetheless, we can relate the two notions as follows:

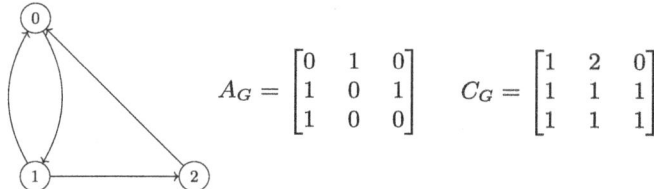

Fig. 2. A rigid graph that is not geometrically rigid, its adjacency matrix A_G and its distance-count matrix C_G.

Theorem 2. *All geometric centrality measures have ties on graphs that are not geometrically rigid. On the other hand, there exists a geometrical centrality measure $\mathfrak{f}$ such that, for all geometrically rigid graphs G, $\mathfrak{f}$ has no ties on G.*

Proof. Define $\mathfrak{f}_G(x)$ to be the rank of x in the lexicographic order of the rows of C_G. Since by definition the rows of C_G are all distinct, the lexicographic order has no ties. On the other hand, if two rows of C_G are the same, the corresponding nodes will have the same value for all geometric centralities (because, by definition, geometric centralities are a function of the distance-count function of the node). □

Whether a similar statement can be proven for *linear* centralities is unclear; what we can prove is a weaker property:

Theorem 3. *For every n, there exists a choice of coefficients $\mathbf{a} \in \mathbf{R}^\mathbf{N}$ such that, for all graphs G with at most n nodes, if G is geometrically rigid then $\mathfrak{L}_G^\mathbf{a}$ has no ties on G.*

Proof. Define $\mathbf{a} = (n^0, n^1, n^2, \ldots, n^{n-1}, 0, 0, \ldots)^T$. Then $\mathfrak{L}_G^\mathbf{a}(i)$ is the base-n encoding of the i-th row of C_G. If G is geometrically rigid, its rows are all distinct, hence $\mathfrak{L}_G^\mathbf{a}$ has no ties. □

6 Distinguishable Coefficients

A natural question we may want to ask is the following: how does $\mathfrak{L}_G^\mathbf{a}$ change as a function of $\mathbf{a}$? Is it possible that using different coefficients we end up defining equivalent linear centralities? The short answer is no, unless the coefficients differ only by a positive multiplicative constant. Let us provide a specific definition for this case.

Definition 8 (Proportionality). *We say that* $\mathbf{a}, \mathbf{b} \in \mathbf{R}^\mathbf{N}$ *are proportional iff there is a* $\lambda > 0$ *such that*

$$b_k = \lambda \cdot a_k \text{ for all } k > 0. \tag{1}$$

Observe that in the definition the 0-th coefficient has no role. In the rest of this section, we can safely assume that the 0-th coefficient of all sequences is 0. Some observations are in order:

- if **a** and **b** are proportional then for all $k > 0$, either a_k and b_k are both zero, or they are both non-zero (and have the same sign);
- no sequence is proportional to $\mathbf{0} = (0, 0, 0, \dots)$, except $\mathbf{0}$ itself;
- if **a** and **b** are proportional, and $\mathbf{a} \neq \mathbf{0}$, then the constant $\lambda > 0$ of Definition 8 is unique;
- if **a** and **b** are not proportional, one of the following statements holds: (a) there is some $k > 0$ such that $a_k b_k = 0$ and $a_k \neq b_k$; (b) for all $i > 0$, $a_i = 0 \iff b_i = 0$ and the minimum index h such that $a_h \neq 0$ satisfies $b_h = \lambda a_h$ for some $\lambda < 0$; (c) for all $i > 0$, $a_i = 0 \iff b_i = 0$ and there exists an index $k > 1$ and $\lambda > 0$ such that $b_i = \lambda \cdot a_i$ for all $0 < i < k$, but $b_k \neq \lambda \cdot a_k$, and moreover $a_h \neq 0$ for some $0 < h < k$.

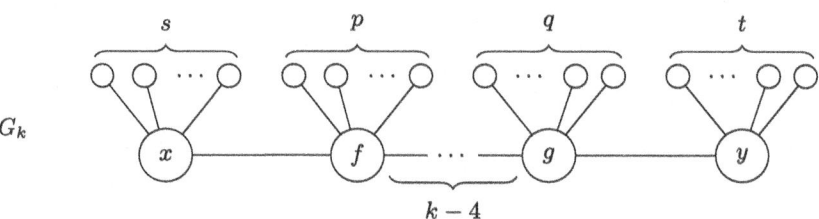

Fig. 3. An undirected graph G_k ($k \geq 4$) used to distinguish two non-proportional sequences. The horizontal dotted line at the bottom represents a path between f and g passing through $k - 4$ vertices.

Based on these observations, we are ready to prove the following:

Theorem 4. *For* $\mathbf{a}, \mathbf{b} \in \mathbf{R}^\mathbf{N}$ *the following holds:*

1. *if* **a** *and* **b** *are proportional, then* $\mathfrak{L}^\mathbf{a}$ *and* $\mathfrak{L}^\mathbf{b}$ *are equivalent centralities;*
2. *if* **a** *and* **b** *are not proportional, then there exists a graph G on which* $\mathfrak{L}^\mathbf{a}_G$ *and* $\mathfrak{L}^\mathbf{b}_G$ *do not agree; in other words, there is a graph G and two nodes* $x, y \in V$ *such that* $\mathfrak{L}^\mathbf{a}_G(x) \geq \mathfrak{L}^\mathbf{a}_G(y)$ *but* $\mathfrak{L}^\mathbf{b}_G(x) < \mathfrak{L}^\mathbf{b}_G(y)$.

Proof. Given a graph $G = (V, E)$, if **a** and **b** are proportional, then $\mathfrak{L}^\mathbf{b}_G(x) = \lambda \cdot \mathfrak{L}^\mathbf{a}_G(x)$ for all $x \in V$, and it is easy to see that in such a case the two centralities are equivalent.

Let us now assume that **a** and **b** are not proportional, and look at the last observation before this theorem. If we are in case (a), and k is the first index at which either sequence is zero and the other is not, any bidirectional path of length k proves the statement: just let x and y the two consecutive nodes at one extreme end of the path. The same path can be used for case (b), using a path of length h.

We are left to show the statement for two non-proportional **a** and **b** both different from **0** that satisfy the condition (c) of the last observation. We will first assume that there exists an index $k > 1$ such that

$$(a_1 - a_k)(b_2 - b_{k-1}) \neq (a_2 - a_{k-1})(b_1 - b_k), \qquad (2)$$

and the cases where this does not hold will be discussed later; note that necessarily $k \geq 4$.

Consider the smallest k satisfying (2), and the undirected graph G_k in Fig. 3, with $n = k+s+p+q+t$ vertices. The linear centralities of x and y with respect to **a** and **b** in this graph are

$$\mathcal{L}^{\mathbf{a}}_{G_k}(x) = e_a + a_1 s + a_2 p + a_{k-1} q + a_k t$$
$$\mathcal{L}^{\mathbf{b}}_{G_k}(x) = e_b + b_1 s + b_2 p + b_{k-1} q + b_k t$$
$$\mathcal{L}^{\mathbf{a}}_{G_k}(y) = e_a + a_1 t + a_2 q + a_{k-1} p + a_k s$$
$$\mathcal{L}^{\mathbf{b}}_{G_k}(y) = e_b + b_1 t + b_2 q + b_{k-1} p + b_k s,$$

where $e_a = a_1 + a_2 + \cdots + a_{k-1}$ and $e_b = b_1 + b_2 + \cdots + b_{k-1}$. Then, the linear centralities $\mathcal{L}^{\mathbf{a}}_G$ and $\mathcal{L}^{\mathbf{b}}_G$ do not agree on G_k if there is an integer nonnegative solution (s, p, q, t) to the following system of inequalities:

$$\begin{cases} e_a + a_1 s + a_2 p + a_{k-1} q + a_k t \geq e_a + a_1 t + a_2 q + a_{k-1} p + a_k s \\ e_b + b_1 s + b_2 p + b_{k-1} q + b_k t < e_b + b_1 t + b_2 q + b_{k-1} p + b_k s, \end{cases}$$

which is equivalent to

$$\begin{cases} \alpha(s - t) + \alpha'(p - q) \geq 0 \\ \beta(s - t) + \beta'(p - q) < 0, \end{cases} \qquad (3)$$

where we have set $\alpha = a_1 - a_k$, $\beta = b_1 - b_k$, $\alpha' = a_2 - a_{k-1}$ and $\beta' = b_2 - b_{k-1}$. The two inequalities in (3) define two halfplanes (in the coordinate system $X = s - t$ and $Y = p - q$), whose intersection is unbounded and nonempty if $\alpha\beta' \neq \alpha'\beta$, which is exactly the condition (2). As a result, the system (3) has an integral solution (X, Y), whence nonnegative s, t, p, q can be always found.

We are left to consider the case in which there is no k satisfying condition (2), but the condition (c) of the observation holds. Then, we can apply the same line of reasoning as above to the graphs $\widehat{G}_k$ in Fig. 4. □

It is a noteworthy fact that all the graphs used to prove the second item of the statement above are undirected, and also connected as long as the condition (2) holds for at least an index k.

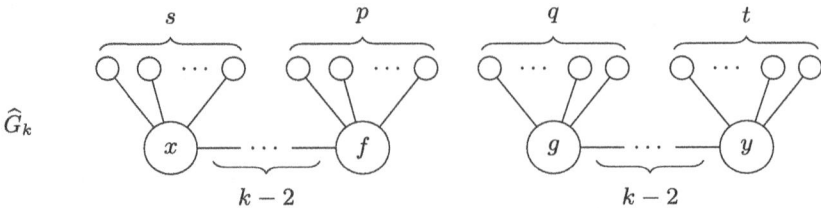

Fig. 4. The graph used to distinguish non-proportional sequences when the condition (2) in the proof of Theorem 4 does not hold. The horizontal dotted lines at the bottom represent paths passing through $k-2$ vertices.

7 Graphs with Many Representable Permutations

The previous section aimed at discussing when and on which graphs two sequences of coefficients yield different centralities. We now ask a dual question: given a graph G, which permutations of its nodes are induced by some linear centrality? In principle, we may have graphs where all linear centralities agree, regardless of how you choose the coefficients: the fact that there is a graph on which each pair of non-proportional centralities can be distinguished does not rule out the possibility that *on some graph* they may all agree.

On the other hand, it may be possible to have graphs that allow for many different permutations of its nodes, depending on the choice of coefficients. To avoid the complications introduced by nodes having the same centrality value, let us stick to the case with no ties.

Definition 9 (Representable permutation). *We say that a permutation $\pi \in S_n$ is representable by a graph G with n nodes iff there is a choice of coefficients $\mathbf{a} \in \mathbf{R}^\mathbf{N}$ such that $\mathcal{L}_G^\mathbf{a}$ has no ties and induces the permutation π.*

Of course, we may classify graphs depending on how many permutations they can represent. More precisely, let us define the *representativeness* of a graph G with n nodes as follows:

$$R(G) = \frac{|\{\pi \in S_n \mid \pi \text{ is representable by } G\}|}{n!} \leq 1.$$

This ratio quantifies how expressive linear centralities are over a rigid graph G. If $R(G) = 1$, linear centralities cannot be less expressible than any other centrality: by choosing coefficients suitably, we can induce any permutation of the nodes, hence achieving what any other centrality can achieve. On the other hand, of course, if $R(G) < 1$ there are permutations that cannot be obtained using a linear centrality, although they may be obtained by some other centrality. In the rest of this section we show that at least in some cases linear centralities are maximally expressive (i.e., they have maximum representativeness).

7.1 Rouché-Capelli Theorem for the Non-connected Case

Given a graph G of n nodes, and a vector $\mathbf{v} \in \mathbf{R}^n$, determining if there exists an $\mathbf{a} \in \mathbf{R}^N$ such that $\mathfrak{L}_G^{\mathbf{a}}(i) = v_i$ for all $i \in [n]$ can be seen as trying to find a solution $\mathbf{a}$ for the following linear system of equations:

$$C_G \cdot \mathbf{a} = \mathbf{v}. \qquad (4)$$

This is a system of n linear equalities with n unknowns, hence by the Rouché-Capelli theorem [17], it has a solution if and only if

$$\operatorname{Rank}(C_G) = \operatorname{Rank}(C_G|\mathbf{v}), \qquad (5)$$

where $\operatorname{Rank}(A)$ is the rank of matrix A, and $A|\mathbf{x}$ is the matrix A augmented with the column $\mathbf{x}$.

Now, if $\operatorname{Rank}(C_G) = n$, Eq. (5) always holds, so if we find a graph G whose distance-count matrix has full rank, then the graph has representativeness 1. Much more than this, in fact: not only can we find linear centralities that rank the nodes in any order we please, but we may even choose the *actual centrality values* for each single node. Such a graph exists:

Theorem 5. *For every $n \geq 3$, the graph G_n in Fig. 5 has representativeness $R(G_n) = 1$.*

Proof. The graph in Fig. 5 has the following distance-count matrix

$$C = \begin{bmatrix} 1 & 0 & 0 & 0 & 0 & \ldots & 0 & 0 & 0 \\ 1 & n-1 & 0 & 0 & 0 & \ldots & 0 & 0 & 0 \\ 1 & 1 & n-2 & 0 & 0 & \ldots & 0 & 0 & 0 \\ 1 & 1 & 1 & n-3 & 0 & \ldots & 0 & 0 & 0 \\ \vdots & \vdots & \vdots & \vdots & \vdots & \ddots & \vdots & \vdots & \vdots \\ 1 & 1 & 1 & 1 & 1 & \ldots & 1 & 2 & 0 \\ 1 & 1 & 1 & 1 & 1 & \ldots & 1 & 1 & 1 \end{bmatrix}$$

which is easily seen to have rank n (it is lower-triangular with no zeros on its diagonal). □

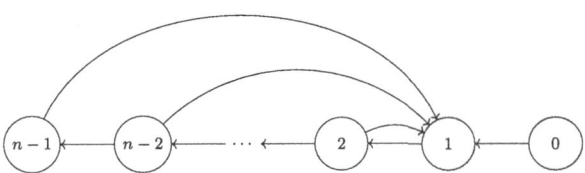

Fig. 5. A graph G_n with representativeness 1.

7.2 Farkas' Lemma for the General Case

The main drawback of the solution in Fig. 5 is that the graph G_n is not strongly connected, albeit almost so. So, we still wonder if a *strongly connected* graph can be permuted in all possible ways by linear centralities.

To tackle this problem (or, more in general, to have a tool to decide if a given permutation is representable by a graph), we shall use the following well-known result [9,10]:

Lemma 1 (Farkas' Lemma). *Given $A \in \mathbf{R}^{m \times n}$ and $\mathbf{b} \in \mathbf{R}^m$, the following statements are equivalent:*

1. *there exists $\mathbf{x} \in \mathbf{R}^n$ such that $A\mathbf{x} \leq \mathbf{b}$;*
2. *there exists no $\mathbf{y} \in \mathbf{R}^m$ such that $A^T\mathbf{y} = \mathbf{0}$, $\mathbf{y} \geq \mathbf{0}$ and $\mathbf{b}^T\mathbf{y} < 0$.*

A consequence of Farkas' Lemma is the following:

Corollary 1. *Given $A \in \mathbf{R}^{m \times n}$, the following statements are equivalent:*

1. *there exists $\mathbf{x} \in \mathbf{R}^n$ such that $A\mathbf{x} < \mathbf{0}$;*
2. *there exists no $\mathbf{y} \in \mathbf{R}^m$ such that $A^T\mathbf{y} = \mathbf{0}$, $\mathbf{y} \geq \mathbf{0}$ and $\mathbf{y} \neq \mathbf{0}$.*

We shall use Corollary 1 to determine if a given graph can represent a permutation. Define the matrix $\Delta \in \mathbf{R}^{(n-1) \times n}$ as follows

$$\Delta = \begin{bmatrix} -1 & 1 & 0 & 0 & \ldots & 0 & 0 & 0 \\ 0 & -1 & 1 & 0 & \ldots & 0 & 0 & 0 \\ 0 & 0 & -1 & 1 & \ldots & 0 & 0 & 0 \\ \vdots & \vdots & \vdots & \vdots & \ddots & \vdots & \vdots & \vdots \\ 0 & 0 & 0 & 0 & \ldots & -1 & 1 & 0 \\ 0 & 0 & 0 & 0 & \ldots & 0 & -1 & 1 \end{bmatrix}$$

For every vector $\mathbf{v} \in \mathbf{R}^n$, the vector is monotonically decreasing (i.e., $v_0 > v_1 > v_2 > \cdots > v_{n-1}$) if and only if $\Delta \mathbf{v} < \mathbf{0}$, because the first inequality is $-v_0 + v_1 < 0$, that is, $v_0 > v_1$, and so on. From this simple observation, we can provide a necessary and sufficient condition for a permutation to be representable.

Theorem 6. *The permutation π is representable by the graph G if and only if there does not exist $\mathbf{y} \in \mathbf{R}^{n-1}$ such that*

$$(\Delta R_\pi C_G)^T \mathbf{y} = 0 \tag{6}$$

with $\mathbf{y} \geq \mathbf{0}$ and $\mathbf{y} \neq \mathbf{0}$, where R_π is the permutation matrix of π (see Sect. 2).

Proof. This is obtained applying Corollary 1 with $A = \Delta R_\pi C_G$. □

In other words, to see if π is representable by G we can look at the system (6) of n equations with $n-1$ indeterminates: π is representable if and only if no non-negative non-trivial solution exists for the system. It is useful to rewrite the i-th element of the LHS of the system (6) in a more intelligible form:

$$\left((\Delta R_\pi C)^T \mathbf{y}\right)_i = \sum_{j<n-1} (\Delta R_\pi C)^T_{i,j} y_j = \sum_{j<n-1} (\Delta R_\pi C)_{j,i} y_j$$
$$= \sum_{j<n-1} \sum_{h_1,h_2} \Delta_{j,h_1}(R_\pi)_{h_1,h_2} C_{h_2,i} y_j$$

where the variables h_1, h_2 range over $[n]$. Now, recalling the definition of Δ and R_π, we obtain that (6) can be rewritten as:

$$\sum_{j<n-1} \left(c_{\pi(j+1),i} - c_{\pi(j),i}\right) y_j = 0 \qquad \forall i \in [n]. \tag{7}$$

If we let $\rho = \pi^{-1}$, we can re-factor these equations as:

$$\sum_{h:\ 0<\rho(h)} c_{h,i} y_{\rho(h)-1} - \sum_{h:\ \rho(h)<n-1} c_{h,i} y_{\rho(h)} = 0 \qquad \forall i \in [n]. \tag{8}$$

A useful observation is that, if we look at a column i of matrix C having only one single non-zero entry $c_{h,i}$, then the condition (8) becomes

$$y_{\rho(h)} = y_{\rho(h)-1}$$

if $0 < \rho(h) < n-1$; if, instead, $\rho(h) = 0$ ($\rho(h) = n-1$, respectively) then the condition implies $y_0 = 0$ ($y_{n-2} = 0$, resp.).

Note that any linear combination of columns of C yields a valid constraint, too. In other words, we can freely take linear combinations of the columns of C and any $\mathbf{y}$ satisfying the constraints of (8) must also satisfy the new constraints.

7.3 Representativeness of Strongly Connected Graphs

As a first attempt at providing a sequence of strongly connected graphs with large representativeness, we have the following:

Lemma 2. *For $n > 3$, consider the graph G_n in Fig. 1, and let $\pi \in S_n$ and $\rho = \pi^{-1}$. The following two statements are equivalent:*

1. *$\rho(0) < \rho(n-1) < \rho(1)$ or $\rho(1) < \rho(n-1) < \rho(0)$;*
2. *the system (6) has no non-negative non-trivial solution.*

Proof. For the sake of space, we omit the proof that is based on an application of (8) and the observations following it. □

It is easy to see (by induction on n) that there are $n!/3$ permutations π satisfying the condition of the above theorem.

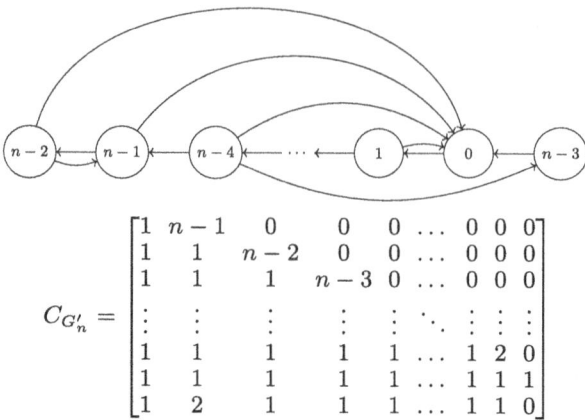

Fig. 6. The graph G'_n with $R(G'_n) = (n-3)/(n-1)$ and its distance-count matrix $C_{G'_n}$.

Corollary 2. *For every $n > 3$, the graph G_n in Fig. 1 is strongly connected and $R(G_n) = 1/3$.*

The idea of representable permutations, besides being useful to estimate expressiveness of linear centralities on a given graph, has another opposite application that we may call *robustness*. Suppose that you know that an adversary will rank the nodes of your graph using *some unknown* linear centrality (it can be closeness, harmonic, or anything else), and suppose that you want to be sure, say, that a certain node (or set of nodes) is ranked higher than another node (or set of nodes). If you can fiddle a bit with the arcs of the graph, you may want to try to impose a structure which guarantees that only the permutations that you like are representable, thus ruling out the possibility that the event you dislike happens, *whatever linear centrality will be applied by the adversary*. The usage of equation (8) is a very powerful tool to guarantee this kind of properties.

A further quest in the direction of finding graphs with large representativeness led us to consider the graph G'_n of Fig. 6. We have experimental results suggesting that $R(G'_n) = (n-3)/(n-1)$, which would imply

$$\lim_{n \to \infty} R(G'_n) = 1,$$

but so far we were unable to prove this result formally. This statement would mean that we have a class of strongly connected graphs that are maximally representative, although only asymptotically so. Whether we can find a family that is ultimately maximally representative remains unknown.

8 Conclusion and Open Problems

This paper presented and studied for the first time the class of linear geometric centralities, providing general properties of expressivity and robustness. Here is a list of problems this paper suggests but that we were unable to solve:

- Does there exist a single linear centrality that has no ties on all geometrically rigid graphs? (In Theorem 3 we solved this problem only for the bounded case)
- Is it possible to find strongly connected graphs to distinguish pairs of non-proportional coefficients, in the cases not covered by the proof of Theorem 4?
- Can we prove formally that $R(G'_n) = (n-3)/(n-1)$ (last paragraph of Sect. 7.3)?
- Is it possible to find some family G'''_n of strongly connected graphs such that $R(G'''_n) = 1$, at least for sufficiently large n?

Acknowledgements. This work was supported in part by project SERICS (PE00000014) under the NRRP MUR program funded by the EU - NGEU. Views and opinions expressed are however those of the authors only and do not necessarily reflect those of the European Union or the Italian MUR. Neither the European Union nor the Italian MUR can be held responsible for them.

References

1. Anthonisse, J.M.: The rush in a directed graph. Technical report BN 9/71, Mathematical Centre, Amsterdam (1971)
2. Babai, L., Luks, E.M.: Canonical labeling of graphs. In: Proceedings of the Fifteenth Annual ACM Symposium on Theory of Computing, pp. 171–183 (1983)
3. Bavelas, A.: A mathematical model for group structures. Hum. Organ. **7**, 16–30 (1948)
4. Bavelas, A.: Communication patterns in task-oriented groups. J. Acoust. Soc. Am. **22**(6), 725–730 (1950)
5. Boldi, P., D'Ascenzo, D., Furia, F., Vigna, S.: Score and rank semi-monotonicity for closeness, betweenness, and distance-decay centralities. Soc. Netw. Anal. Min. **14**(1), 183 (2024)
6. Boldi, P., Vigna, S.: Axioms for centrality. Internet Math. **10**(3–4), 222–262 (2014)
7. Bollobás, B.: Modern Graph Theory. Graduate Texts in Mathematics, vol. 184. Springer (1998)
8. Brandes, U., Erlebach, T.: Network Analysis: Methodological Foundations (Lecture Notes in Computer Science). Lecture Notes in Computer Science, vol. 3418 . Springer (2005)
9. Farkas, J.: Theorie der einfachen ungleichungen. J. für die reine und angewandte Mathe. (Crelles J.) **1902**(124), 1–27 (1902)
10. Gale, D., Kuhn, H.W., Tucker, A.W.: Linear programming and the theory of games. Act. Anal. Prod. Allocat. **13**, 317–335 (1951)
11. Katz, L.: A new status index derived from sociometric analysis. Psychometrika **18**(1), 39–43 (1953)

12. Kishi, G.: On centrality functions of a graph. In: Saito, N., Nishizeki, T. (eds.) Graph Theory and Algorithms. LNCS, vol. 108, pp. 45–52. Springer, Heidelberg (1981). https://doi.org/10.1007/3-540-10704-5_5
13. Kishi, G., Takeuchi, M.: A type of centrality functions for a directed graph. Electron. Commun. Jpn. (Part I: Commun.) **66**(5), 38–47 (1983)
14. Lin, N.: Foundations of Social Research. McGraw-Hill, New York (1976)
15. Sabidussi, G.: The centrality index of a graph. Psychometrika **31**(4), 581–603 (1966)
16. Schoch, D., Brandes, U.: Re-conceptualizing centrality in social networks. Eur. J. Appl. Math. **27**(6), 971–985 (2016)
17. Shafarevich, I.R., Remizov, A.O.: Linear Algebra and Geometry. Springer (2012)
18. Skibski, O., Sosnowska, J.: Axioms for distance-based centralities. In: Proceedings of the AAAI Conference on Artificial Intelligence, vol. 32, no. 1 (2018)

Analysis and Predictability of Centrality Measures in Competition Networks

Anthony Bonato[(✉)] and Mariam Walaa

Department of Mathematics, Toronto Metropolitan University, Toronto, ON, Canada
abonato@torontomu.ca

Abstract. The Common Out-Neighbor (or CON) score quantifies shared influence through outgoing links in competitive contexts. A dynamic analysis of competition networks reveals the CON score as a powerful predictor of node rankings. Defined in first-order and second-order forms, the CON score captures both direct and indirect competitive interactions, offering a comprehensive metric for evaluating node influence. Using datasets from Survivor, Chess.com, and Dota 2 online gaming competitions, directed competition networks are constructed, and the dynamic CON score is integrated into supervised machine learning models. Empirical results show that the CON score consistently outperforms traditional centrality measures such as PageRank, closeness, and betweenness centrality in classification tasks.

By integrating dynamic centrality measures with machine learning, our proposed methodology accurately predicts outcomes in competition networks. The findings underline the CON score's robustness as a feature in node classification, offering a significant advancement in understanding and analyzing competitive interactions.

1 Introduction

Centrality measures were used to study real-life networks since the early days of modern network science [1,7,12,13] and, more recently, as predictors in machine learning problems [8,14,16]. They are crucial for identifying actors who are influential in a complex network and are often considered to be local metrics.

Competition networks model adversarial relationships between actors. Examples of competition networks constructed using real-life data include e-sports win networks, and voting networks for reality competition shows. In [4], the authors develop a tool to measure the correlation between known winners and their membership in strong alliances within competition networks from reality television shows Survivor and Big Brother. Alliance strength in these competitions is quantified by analyzing the edge densities of known alliances.

In [5], the authors introduce the CON score as a predictor of influential actors in competition networks. More broadly, they proposed the *Dynamic Competition Hypothesis* (or DCH), which suggests that leaders in adversarial networks

---

A. Bonato—Supported by an NSERC Discovery Grant.

exhibit a high CON score, high closeness, low in-degree, and high out-degree. In [6], the concept of *low-key leaders* was introduced, defined as nodes that maintain influence despite low centrality (from low PageRank). These leaders are identified by contrasting their high CON scores with their PageRank values [7], demonstrating that influence in competition networks can arise from factors beyond traditional centrality measures.

Our objective in this paper is to analyze the centrality of competition networks constructed from game-based datasets collected from online sources and to build network-based machine learning models capable of accurately predicting known outcomes of these games. We claim that linking centrality measures to adversarial interactions and analyzing them over time provides an effective method for ranking nodes in competition networks, providing a high correlation with ground truth data. We demonstrate that the CON score is a highly effective predictive feature, often outperforming traditional centrality measures such as PageRank, closeness, and betweenness centrality.

The paper is organized as follows. Section 2 introduces the first-order and second-order CON scores and provides definitions of other centrality measures used in this study. Section 3 describes the network datasets and explains the process of feature generation for building machine learning models using the CON score and other centrality measures. Finally, we conclude with a summary of our findings and discuss open problems for future research.

We consider weighted, directed graphs (or *digraphs*) with directed edges in the paper. The *in-neighbors* and *out-neighbors* of node v are denoted by $N^-(v)$ and $N^+(v)$, respectively. The in- and out-degree of v is denoted by $\deg^-(v)$ and $\deg^+(v)$, respectively. Additional background on graph theory may be found in [18], and more background on complex networks may be found in [3].

2 The CON Score

High centrality is intuitively one of the key characteristics of an influential node in a competition-based network. A strong player is expected to engage with many others in the network and emerge victorious in these interactions. In this section, we formalize these ideas by defining competition networks and their dynamic extensions.

A *competition network* is a directed graph $G = (V, E)$, where each node $u \in V$ represents a competitor, and a directed edge $(u, v) \in E$ exists if u has defeated v at least once in a competition. This structure captures the competitive interactions and relationships among actors in the network. A *dynamic competition network* extends this concept across multiple rounds or time-steps. It is represented as a sequence of graphs $(G_t : 1 \leq t \leq T)$, where each $G_t = (V, E_t)$ describes the competition network at time step t. The set of nodes V (which represent the competitors) remains fixed across time, while $E_t \subseteq V \times V$ denotes the directed edges (or competitions) occurring at time t.

To quantify the competitive interactions between two nodes, we define the *Common Out-Neighbor* (or CON) score of u and v as the number of nodes to which both u and v have directed edges. Specifically, $\text{CON}(u,v)$ measures the number of common competitors that u and v have defeated. For a detailed mathematical definition of $\text{CON}(u,v)$, refer to [5].

The *CON score of u* quantifies the number of common out-neighbors a node shares with the rest of the network. We compute the CON score for every node in a competition network and use it to rank nodes based on competitiveness. Let v be an arbitrary node in graph G, and let $\mathbf{A}$ be the adjacency matrix of G. To account for first-order neighbors, we consider $\mathbf{A}[i,j]$, the (i,j) entry of $\mathbf{A}$, which represents the number of directed edges from i to j, or the number of victories by actor i against actor j. This gives:

$$\text{CON}_1(v) = \sum_{\substack{u,x \in V(G),\\ u \neq v}} \min(\mathbf{A}[v,x], \mathbf{A}[u,x]).$$

This formula applies to both weighted and unweighted graphs. We next introduce a new measure relevant to competition networks: the *second-order CON score*, which considers the number of common out-neighbors up to distance two in the competition network. The purpose of the second-order CON score is to capture competitors of competitors, extending beyond direct competition. For example, if player u competes with player w, and player v competes with player x, who in turn competes with w, the first-order CON score does not capture this (u,w)-(v,w) CON pair, but the second-order CON score does. For an illustration of this, see Fig. 1.

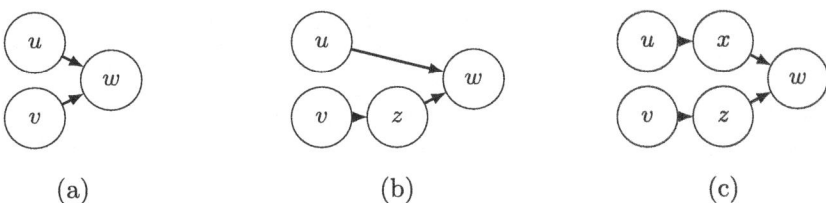

Fig. 1. Three scenarios displaying direct and indirect competition in a network. In (a), both u and v have direct competition with w, whereas in (b), only v has indirect competition with w, and in (c), both u and v have indirect competition with w.

To account for second-order neighbors, which correspond to all distance-two paths, we compute the square of the adjacency matrix. Summing both distance-one and distance-two paths gives:

$$\mathbf{A}_2[i,j] = \mathbf{A}[i,j] + \mathbf{A}^2[i,j].$$

For the second-order CON score, we have:

$$\text{CON}_2(v) = \sum_{\substack{u,x \in V(G),\\ u \neq v}} \min(\mathbf{A}_2[v,x], \mathbf{A}_2[u,x]).$$

To evaluate the effectiveness of the second-order CON score, we compare it with well-established centrality measures used for analyzing node importance. The *closeness centrality* [1] of a node v in G is defined as:

$$C(v) = \left(\sum_{u \in V(G) \setminus \{v\}} d(v, u) \right)^{-1},$$

where $d(v, u)$ is the shortest path distance from v to u, if such a path exists. The *betweenness centrality* [12] is defined as:

$$B(v) = \sum_{x, y \in V(G) \setminus \{v\}} \frac{\sigma_{xy}(v)}{\sigma_{xy}},$$

where $\sigma_{xy}(v)$ is the number of shortest directed paths between x and y that pass through v, and σ_{xy} is the total number of shortest directed paths between x and y.

A final, well-known centrality measure we consider is PageRank [7], albeit applied to the digraph formed by reversing the edges of the network. For more on PageRank, see [3].

3 Experimental Design and Methods

We consider a variety of game data to construct our competition networks. We start with a dataset for *Survivor*, a famous reality competition show that began in the U.S. in 2000. The American *Survivor* competitions are structured consistently across all independent seasons. In each season, approximately 16 players are divided into tribes and must compete to win the show's grand prize. Individuals form latent, shifting alliances within tribes to eliminate weak or threatening tribe members. Despite the team-based competition, each castaway must ultimately fight for their chance at the grand prize. More details about the competition are provided in [4]. In the *Survivor* network, individual players are represented as nodes, and votes between players form directed edges.

The second dataset we work with is from the *Chess.com Titled Tuesday* competitions. These competitions follow FIDE rules and guidelines as outlined in [11]. Players are initially ranked based on strength (rating), FIDE title, and alphabetically. This ranking, which is referred to as the *Initial Order (or IO)*, determines the *Pairing Numbers* (or *PN*). After each round, *scoregroups* are formed, consisting of players with the same score. A *pairing bracket* is considered homogeneous if all players within it have the same score; otherwise, it is classified as heterogeneous. Players within a scoregroup may sometimes have different scores because of *downfloaters* or *upfloaters*, who are moved from their original brackets to facilitate pairings. After each round, players are sorted lexicographically, first by scoregroup and then by IO. Pairings occur once scoregroups and pairing brackets are finalized. In general, the FIDE pairing system ensures that players compete with others of similar skill in the first round and are paired in subsequent rounds based on performance and overall skill. In the Chess.com network, players are represented as nodes and directed edges correspond to match outcomes.

Finally, we use a dataset from *Dota 2*, a Multiplayer Online Battle Arena (or MOBA) game developed in 2013. The original game, Dota, was among the first to establish the MOBA genre, becoming widely popular. Dota 2 has since achieved over 80 million accounts in the past decade. The Dota 2 Professional League matches represent a subset of Dota 2 games that are collected, tracked, and updated weekly on Kaggle, providing detailed insights into matches and team performance. The Dota 2 competition network is constructed by representing each team as a node and connecting them with directed edges that indicate match outcomes, where an edge points from the winning team to the losing team.

Our datasets were selected based on availability and relevance. Experiments conducted on these datasets demonstrate that the CON score can serve as a powerful predictive feature in machine learning models for node classification. We construct the competition networks as follows.

1. *Survivor*: Voting data for 46 seasons of *Survivor* [17] was downloaded from the public GitHub repository, survivoR2py. Two files were primarily used to construct the network: vote_history.csv, which contains castaways and their votes in each episode, and castaways.csv, which contains each castaway's outcome at the end of the season (for example, sole survivor, runner-up, or n-th voted out). After data cleaning, a total of 726 castaways from 46 seasons were extracted, with 3,662 tribal council votes across an average of 12 episodes per season.
2. *Chess.com*: Tournament data from Chess.com's Titled Tuesday webpage [9] was scraped. Two files containing results and pairings for each tournament round were utilized. The tournaments operate in a Swiss format, with players "dropping in" weekly to participate in early or late games over 11 rounds. Each Chess.com player has a Glicko rating that determines their overall ranking in terms of chess skill. The Chess.com dataset we considered had 863 nodes and 16,881 edges.
3. *Dota 2*: Professional league match data was downloaded from Kaggle, and OpenDOTA rankings were retrieved using the OpenDOTA API in Python. Each Dota 2 team has a Glicko rating that determines their rank in terms of Dota 2 skill. Two files were used to construct the network. First, there was main_metadata.csv, which contains details about each match, including team IDs and whether a team was radiant or dire, and second, teams.csv, which contains team names. The Dota 2 dataset we considered had 493 nodes and 2,413 edges.

An important aspect of supervised machine learning modeling is the acquisition of ground truth labels. In our datasets, ground truth labels represent the true ranking of nodes in a competition network (for example, player rankings

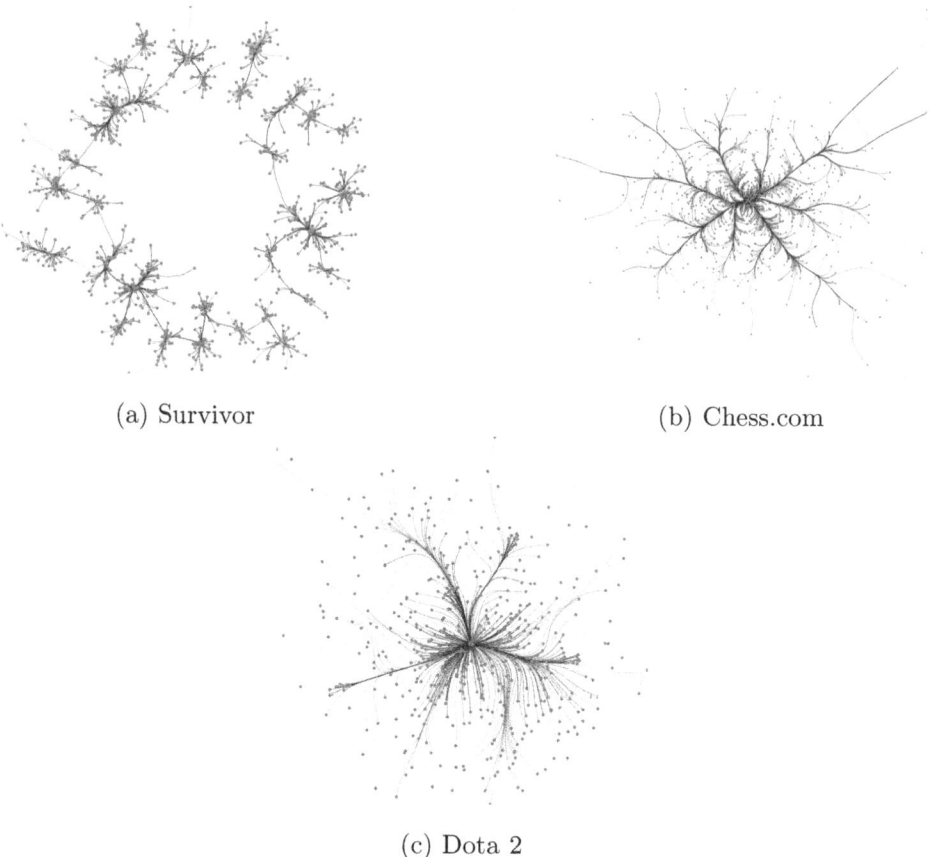

(a) Survivor (b) Chess.com

(c) Dota 2

Fig. 2. Force-directed, edge-bundled networks based on game datasets; see [2]. In (a), smaller disconnected networks organized in a circular structure represent 46 independent seasons of Survivor. In (b), a highly centralized, connected network represents Chess matches between players. In (c), disconnected networks with a dense central core represent top teams playing matches in Dota 2. Both (b) and (c) show low-interaction players/teams on the periphery, with high-interaction players/teams in the core.

or outcomes). These labels are used to evaluate how centrality-based predictions align with actual outcomes. Ground truth rankings are compared with the model's predictions based on centrality measures, allowing for the approximation of node influence. True rankings are determined by accurate, up-to-date Glicko ratings or holistic game outcomes and are not specific to the sample of game data collected. At the end of each season, each Survivor participant has an *outcome* describing whether they are the sole survivor, a runner-up, or the n-th voted out. Similarly, each Chess.com player has a Glicko rating that defines their rank in terms of chess skill, and each Dota 2 team has a Glicko rating that defines their rank in terms of Dota 2 skill.

As shown in Fig. 2, the nature of the competition networks resulting from these games varies, as different factors across the three datasets drive the presence of an edge between two actors. For instance, in the Chess.com data, players are paired with those of similar skill and game outcomes, while in Survivor, players vote off those they think should be removed from the competition. We refer the reader to https://github.com/mariamwalaa/CON-CN for access to the data, models, and further analysis.

The algorithm computes first-order and second-order CON scores in a competition network where nodes represent actors and directed edges represent one actor defeating another in a specific round of the competition. The algorithm iterates over competitions and rounds and independently computes the CON score for each actor in each round. The *Node Ranking Model* (or NRM) for Survivor is as follows. The algorithm for Chess.com and Dota 2 is almost identical, except without the for loop in line 1.

Algorithm: Dynamic CON score in Survivor.
---
1 **foreach** *competition in competitions* **do**
2   **foreach** *round in competition* **do**
3   Generate a graph of the current round
4   Construct adjacency matrix for actors
5   Square adjacency matrix to consider second-order neighbors
6   Calculate 1st and 2nd order CON scores for each actor
7   Store CON scores in a results list
8   **end**
9 **end**
10 **return** results
---

1. *Network Construction*: Competitions that have occurred have associated data posted online. Once data for a given competition and its associated rounds have been collected, a network is constructed by identifying teams, players, and the winners of each game. Four main variables are of interest: competition, round, game/match, players/teams, and time-step. If multiple competitions occur, then the competition-specific networks are amalgamated into one large network.
2. *Feature Generation*: Once a network is constructed, centrality measures are computed for every node at each time step 1 through k, and a feature matrix is generated with n rows representing teams/players and m columns representing centrality features.
3. *Ground Truth Labels*: To train the models, a classification of nodes based on ground truth data must be performed. We use the current rankings of players as well as known outcomes to create these classes. The numeric rankings are categorized into three quantile groups: *the bottom* 10%, *the top* 10%, and the *middle range* consisting of the remaining 80%.
4. *Model Training*: The final step involves first splitting the feature data into training and testing sets, then training machine learning models on data

points associated with the first 80% of time steps and testing them on the data points associated with the remaining 20% of time steps.

We use supervised machine learning models built for classification purposes to predict node importance. These models are well-suited for handling high-dimensional tabular data, which can be derived from network features by transforming them into a structured format. This transformation enables us to leverage the information encoded in networks, allowing us to take advantage of classification models. A sample of this transformed data is presented in Table 1.

Table 1. Dota 2 data sample from Week 39, sorted by CON score. The Dota 2 team corresponding to ID 8629005.0 had a CON score of 1319, a PageRank score of 0.008925, a closeness score of 0.027778, in-degree of 6, and out-degree of 7. Table only shows metrics for one round of competitions.

Team	CON	PageRank	Closeness	In-Degree	Out-Degree
8629005.0	1319.0	0.008925	0.027778	6.0	7.0
8629317.0	1216.0	0.009633	0.025641	5.0	6.0
9426115.0	1212.0	0.006819	0.025641	5.0	6.0
9425660.0	1204.0	0.004711	0.023810	5.0	7.0
9425656.0	1176.0	0.008009	0.025641	5.0	7.0

We start with a decision tree, a simple model that splits data based on a series of hierarchical feature conditions to classify observations. However, as decision trees are prone to overfitting, we also consider a random forest model, which is a robust ensemble of decision trees. To further enhance predictive performance, we also explore gradient boosting and XGBoost [10], which are better at penalizing overfitting by building trees sequentially while optimizing for errors. Finally, we consider a support vector machine, which falls under a different class of machine learning algorithms and classifies data by finding the best hyperplane; that is, the largest separation between classes.

Table 2 provides details about the size of the networks and how long it takes to run the NRM on each of the datasets. The difference in the size of the edge sets is reflected in the time in seconds it takes to run the model. The performance metrics of the machine learning models are in Table 3. We consider the F1 score rather than the accuracy of each dataset since the classes are imbalanced. We see that the Dota 2 network has the highest accuracy at 0.845 while the Survivor network has the highest F1 score at 0.788.

To determine which centrality features are most important for the top-performing machine learning model in each dataset, we use tree-based feature importance, which measures the *mean decrease in Gini impurity* (or MDI) as

$$\text{MDI}(f_i) = \sum_{t \in T_i} p(t) \Delta I_t,$$

Table 2. Graph descriptions for Survivor, Chess.com, and Dota 2 datasets. The Dota 2 network has the highest diameter of 10 as well as the lowest sparsity of 0.0064, indicating that it is the most spread-out, dense network, while the Survivor network has the highest number of weakly connected components (or WCCs), a total of 46.

Metric	Survivor	Chess.com	Dota 2
# Nodes	806	863	493
# Edges	3,662	15,881	2,413
# Rounds	12	18	8
# Competitions	46	1	1
# Labels	152/455/152	87/690/86	50/393/50
Connected	No	Yes	No
# WCC	46	1	39
# SCC	90	152	199
Sparsity	0.0064	0.0214	0.0081
Diameter	3	4	10
Runtime	1.5 s	635 s	22 s

Table 3. Performance metrics for Survivor, Chess.com, and Dota 2 datasets across various models: support vector machine, random forest, XGBoost, gradient boosting, and decision trees. The highest score for each metric is highlighted in bold.

Model	Survivor				Chess.com				Dota 2			
	Acc.	Prec.	Rec.	F1	Acc.	Prec.	Rec.	F1	Acc.	Prec.	Rec.	F1
SVM	**0.816**	**0.780**	**0.807**	**0.788**	0.799	0.266	0.333	0.296	0.608	0.570	0.662	0.549
RF	0.750	0.705	0.697	0.700	**0.811**	0.472	**0.450**	0.452	0.770	0.645	0.594	0.606
XGB	0.750	0.713	0.695	0.701	0.803	**0.602**	0.436	0.461	0.791	0.628	0.602	0.612
GB	0.728	0.685	0.659	0.666	0.788	0.423	0.429	0.422	**0.845**	**0.714**	**0.663**	**0.686**
DT	0.662	0.607	0.566	0.578	0.699	0.446	0.493	**0.462**	0.777	0.615	0.635	0.620

where T_i is the set of all tree nodes in the forest where f_i is used for splitting, $p(t)$ is the proportion of samples reaching node t (node t's weight in the tree), and ΔI_t is the impurity decrease achieved at node t. The larger the MDI, the more important a feature is deemed to the model. As shown in Fig. 3, the feature importance of the CON score exceeds that of the remaining centrality measures in the Dota 2 dataset. Specifically, we see that there is an uptick in importance corresponding to week 39, which suggests that CON scores of teams who played in week 39 were crucial for predicting the class of teams.

To assess the relationship between centrality metrics and ground truth labels (such as team rankings or player performance), we rank-order the centrality measures and display them as a line chart. A matrix of line charts for the Chess.com dataset is shown in Fig. 4. First, we apply a re-scaling technique known as *unity-*

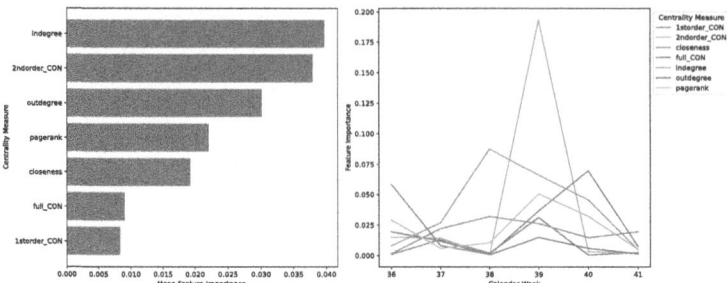

Fig. 3. On the left, a bar chart of Dota 2 mean feature importance (y-axis) versus time-step (x-axis) shows that the CON score has a higher mean feature importance than other centrality measures across all time-steps. On the right, a line chart shows mean importance at each time step, with a jump in CON mean importance at week 39.

based normalization, as described in [6]. Given data points $X_1, X_2, \ldots, X_n$, with minimum value $X_{\min}$ and maximum value $X_{\max}$ where $1 \leq i \leq n$, the normalized value $X_{i,\text{norm}}$ is calculated as

$$X_{i,\text{norm}} = \frac{X_i - X_{\min}}{X_{\max} - X_{\min}}.$$

This scaling method ensures that all normalized values $X_{i,\text{norm}}$ lie within the range $[0, 1]$. For consistency, we apply this normalization method to all centrality measures. In addition to scaling, a 50-point moving average is applied to smooth the data and better visualize the trend.

In Table 4, Spearman's coefficient is used to measure the correlation between the predicted rankings from centrality-based measures and actual ground truth labels. The results show strong correlations, suggesting that centrality features can effectively approximate the true influence of nodes in competition networks. As shown in Fig. 4 and Table 4, there is a high correlation between the CON score and common centrality measures, which supports the hypothesis that the CON score can be used as a centrality measure.

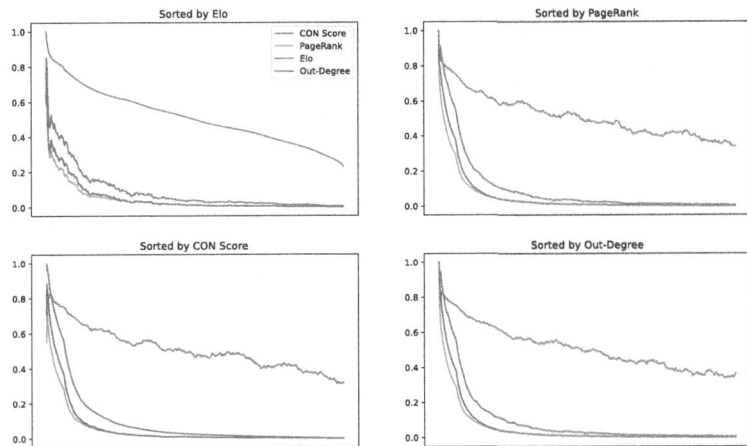

Fig. 4. A 2-by-2 matrix of line charts shows four measures (CON, PageRank, Elo, and Out-Degree) for 861 Chess.com players, each sorted by one measure. Elo is the global ranking system for players based on their chess skills and serves as the ground-truth label. A negative exponential correlation exists between all centralities and the sorted measure of choice. The x-axis represents the 861 players, while the y-axis is the normalized centrality measure.

Table 4. Spearman correlation coefficients (sorted in descending order) between centrality measures and Chess.com rankings show that out-degree has the highest correlation to Chess.com rankings, while closeness has the lowest correlation to Chess.com known rankings.

Metric	Correlation	P-Value
Out-Degree	0.736	6×10^{-148}
CON Score	0.707	2×10^{-131}
PageRank	0.701	2×10^{-128}
Betweenness	0.664	2×10^{-110}
Closeness	0.309	2×10^{-20}

To further validate our findings, we generated a supervised *Uniform Manifold Approximation and Projection* (or UMAP) [15] embedding of the model features to determine whether these features properly distinguish between the three classes of nodes. See Fig. 5. The approach also shows a clear separation between the three classes and localized clustering within the middle class. For example, the middle class in the Survivor dataset consists of three distant clusters, while the Dota 2 middle class has smaller, more dispersed clusters. This aligns with expectations, as the middle class contains the majority of the observations and is likely to exhibit greater variation.

(a) Survivor   (b) Chess   (c) Dota 2

Fig. 5. Plots of the 2-dimensional supervised UMAP embeddings for each actor in the competition networks, where red is the top 10%, green is the bottom 10%, and blue is the middle 10%–90%. (a) shows that Survivor castaways in the middle range are dispersed into three main groups, whereas the top and bottom ranges are more clustered together. (b) shows that Chess.com players in the middle range are more evenly distributed, while the top and bottom ranges are distant from each other. (c) shows that the Dota 2 middle range has many smaller clusters, while the top and bottom ranges are close. The maps show that there is a clear separation between classes with localized clustering. (Color figure online)

4 Discussion and Future Work

We developed a method to dynamically compute the CON score and other common centrality measures for competition networks. This approach enabled the creation of a feature set that accurately predicts the class of a node based on ground truth data. Our analysis confirmed that the CON score effectively seriated nodes, as evidenced by a high Spearman coefficient and strong visual correlation between metrics at the node level.

While we demonstrated aggregate feature importance to establish the strength of the CON score relative to other centrality measures, future work could focus on node-level analysis of feature importance to gain deeper insights. Model performance may also benefit from improved ground truth labels. For instance, in the Dota 2 datasets, team rankings are more closely aligned with the timeline of matches, and additional context about the data collection process could enhance predictions. Beyond games, adversarial networks in fields like biology, such as food webs and dominance hierarchies, offer promising avenues for further exploration.

Future research could apply this algorithm to various adversarial networks and random graph models, including Erdős-Rényi graphs, the preferential attachment model, or random geometric graphs, to assess its broader applicability. Questions about the CON score's nature, such as whether it is a local or global metric, remain to be addressed. Additionally, generalizing the CON score to the k-th order, where k is the diameter or another meaningful parameter, could provide novel insights into network structure and dynamics. By addressing these challenges, we aim to expand the utility of the CON score and contribute to a deeper understanding of complex networks across disciplines.

Disclosure of Interests. The authors have no competing interests to declare that are relevant to the content of this article.

References

1. Bavelas, A.: Communication patterns in task-oriented groups. J. Acoust. Soc. Am. **22**, 725–730 (1950)
2. Bednar, J.A., Crist, J., Cottam, J., Wang, P.: Datashader: revealing the structure of genuinely big data. In: Proceedings of the 15th Python in Science Conference (SciPy 2016) (2016)
3. Bonato, A.: A Course on the Web Graph. American Mathematical Society, Providence (2008)
4. Bonato, A., Eikmeier, N., Gleich, D.F., Malik, R.: Dynamic competition networks: detecting alliances and leaders. In: Proceedings of Algorithms and Models for the Web Graph (WAW 2018) (2018)
5. Bonato, A., Eikmeier, N., Gleich, D.F., Malik, R.: Centrality in dynamic competition networks. In: Proceedings of the International Conference on Complex Networks and Their Applications (2019)
6. Bonato, A., Kapusin, J., Yuan, J.: Winner does not take all: contrasting centrality in adversarial networks. In: Proceedings of the International Conference on Complex Networks and Their Applications (2024)
7. Brin, S., Page, L.: The anatomy of a large-scale hypertextual web search engine. Comput. Netw. **30**, 107–117 (1998)
8. Bucur, D., Holme, P.: Beyond ranking nodes: predicting epidemic outbreak sizes by network centralities. PLoS Comput. Biol. **16**, 1–20 (2020)
9. Chess.com. Live Titled Tuesdays (2024). https://www.chess.com/tournament/live/titled-tuesdays
10. Chen, T., Guestrin, C.: XGBoost: a scalable tree boosting system. In: Proceedings of the 22nd ACM SIGKDD International Conference on Knowledge Discovery and Data Mining (2016)
11. FIDE, C.04.3 FIDE (Dutch) system. FIDE Handbook (2025). https://handbook.fide.com/chapter/C0403202507
12. Freeman, L.C.: Centrality in social networks: conceptual clarification. Soc. Netw. **1**, 215–239 (1979)
13. Katz, L.: A new status index derived from sociometric analysis. Psychometrika **18**, 39–43 (1953)
14. Liu, W., Fang, P., Guo, F., Qiao, Y., Zhu, Y., Wang, H.: Graph-theory-based degree centrality combined with machine learning algorithms can predict response to treatment with antipsychotic medications in patients with first-episode schizophrenia. Dis. Markers (2022)
15. McInnes, L., Healy, J., Melville, J.: UMAP: uniform manifold approximation and projection. J. Open Sour. Softw. **3**, 861 (2018)
16. Morselli, C., Masias, V.H., Crespo, F., Laengle, S.: Predicting sentencing outcomes with centrality measures. Secur. Inform. **2**, 4 (2013)
17. Stiles, P.: survivoR2py. GitHub repository (2023). https://github.com/stiles/survivoR2py
18. West, D.B.: Introduction to Graph Theory, 2nd edn. Prentice Hall (2001)

Modularity of Random Intersection Graphs

Katarzyna Rybarczyk[✉]

Adam Mickiewicz University, Poznań, Poznań, Poland
kryba@amu.edu.pl

Abstract. Modularity was introduced by Newman and Girvan in 2004 and is used as a measure of community structure of networks represented by graphs. In our work we study modularity of the random intersection graph model first considered by Karoński, Scheinerman, and Singer–Cohen in 1999. Since their introduction, random intersection graphs have attracted much attention, mostly due to their application as models of real networks. In our work we determine the range of parameters in which modularity detects the community structure of the random intersection graphs well, as well as give a range of parameters for which there is a community structure present but not revealed by modularity. We also relate modularity of the random intersection graph to the modularity of other known random graph models.

Keywords: Modularity · Random intersection graph · Random graphs · Affiliation networks · Complex networks

1 Introduction

Motivation. Detecting and measuring the presence of community structure in the network are undoubtedly natural motivations for compelling research in the field of computer science. Mainly because identifying communities has a number of practical applications such as identifying the groups of common interests in social networks, classifying fake news, identifying proteins with the same biological functions and many others [22,31]. In this article we concentrate on the notion of measure of the presence of community structure in the network's graph called *modularity*. Modularity was introduced by Newman and Girvan in 2004 [31,32]. It relays on the idea that in the graph with a present community structure the vertex set might be partitioned into subsets in which there are much more internal edges than we would expect by chance (see Definition 1). Modularity is commonly used in community detection algorithms as a quality function judging the performance of the algorithms [22], and also as a central ingredient of such algorithms, for example in the Louvain algorithm [4], Leiden algorithm [40] or Tel-Aviv algorithm [15].

The first theoretical results concerned modularity of deterministic graphs (see for example [1,10,27]). This line of research is still continued, see for example a recent result [26] by Lasoń and Sulkowska. However, in the context of studying large networks, it is crucial to determine the modularity of their natural models – random graphs. Systematic and thorough studies on this topic have been undertaken by McDiarmid and Skerman in a series of articles [28–30]. However, only recently has research begun on the modularity of these random graph models, which are considered to reflect the properties of complex networks. Here we should mention the results about modularity of the random graph on the hyperbolic plane by Chellig, Fountoulakis, and Skerman [12], those concerning the basic preferential attachment graph by Prokhorenkova, Prałat, and Raigorodskii [33,34] and by Rybarczyk and Sulkowska [38], ABCD random graph model by Kamiński et al. [20], and generalisations to hypergraphs [19,21].

It is a plausible that many of the properties of complex networks are related to a known or hidden bipartite structure representing the relationships between network elements and their properties and the fact that alike elements (with common properties) tend to form connections [17]. A theoretical model of a random graph based on a bipartite structure of vertices and attributes has been introduced by Karoński, Scheinerman, and Singer-Cohen in [23] and called the random intersection graph. The model was further generalised by Godehardt and Jaworski [16]. Moreover, some other variants of graphs based on a bipartite structure have been studied since, for example [7–9,41]. In classical random intersection graphs, vertices choose their neighbours based on randomly assigned attributes they possess. It might be interpreted in such a manner that the more the elements of the network are alike, the more probable is that they are connected. For comprehensive surveys on the classical results concerning random intersection graphs we refer the reader to [5,6,39]. It turns out that with well-chosen parameters, random intersection graphs have many properties of complex networks. To mention just a few, random intersection graphs are known for having tunable clustering properties and degrees that in some ranges of parameters have the power law distribution, see for example [6] and references therein. One of the reasons for having these properties is the fact that, unlike in Erdős-Rényi graph, in random intersection graphs edges do not appear independently. We stress here the fact that, in many ranges of parameters, random intersection graphs have large clustering i.e. apparently have some community structure. However still little is known about their modularity.

In our work we concentrate on the modularity of the classical model of the random intersection graph introduced in [23]. However we hope that this work will be followed by a more thorough study of other random intersection graph models.

Notation. Before we give a formal definition of the considered model we introduce some basic notations. We denote by $[n] = \{1, 2, \ldots, n\}$ the set of the first n natural numbers. By $(n)_k = n \cdot (n-1) \cdots (n-k+1)$ we denote the falling factorial. Moreover, in what follows, we use standard asymptotic notations $o(\cdot), O(\cdot), \Omega(\cdot), \Theta(\cdot), \sim, \asymp, \ll, \gg$, as defined in [18]. Also $\omega = \omega_n \to \infty$ will be a function tending slowly to infinity as $n \to \infty$. All limits are taken as $n \to \infty$ and inequalities hold for large n. We say that an event $\mathcal{E}$ holds with high probability if $\Pr(\mathcal{E}) \to 1$ as $n \to \infty$. We use $\mathrm{Bin}(\cdot,\cdot)$ and $\mathrm{Po}(\cdot)$ to denote the binomial and Poisson distribution, respectively. For any graph G we denote by $V(G)$ its vertex set and by $E(G)$ its edge set. Moreover for any subset S of the vertex set, $S \subseteq V(G)$, of a graph G we denote by $\bar{S} = V(G) \setminus S$ its complement. By $e_G(S)$ we denote the number edges in G with both ends in S and by $e_G(S, \bar{S})$ the number of edges with exactly one end in S. In addition by $\mathrm{vol}_G(S) = 2e_G(S) + e_G(S, \bar{S})$ we denote the sum of degrees of vertices from S in G. We omit the subscript G, when it is clear from the context which graph G we have in mind. Moreover, we write $e(G) = e_G(V(G))$ and $\mathrm{vol}(G) = \mathrm{vol}_G(V(G))$.

Model and its Properties. In the binomial random intersection graph $\mathcal{G}(n, m, p)$ introduced in [23] there is a set of vertices $\mathcal{V} = \{v_1, \ldots, v_n\}$ and an auxiliary set of attributes $\mathcal{W} = \{w_1, \ldots, w_m\}$. All vertices v_i, $i \in [n]$, are attributed random subsets $\mathcal{W}(v_i)$, $i \in [n]$, of $\mathcal{W}$ in such a manner that for every v_i, $i \in [n]$, each $w \in \mathcal{W}$ is included in $\mathcal{W}(v_i)$ independently at random with probability p. Two vertices v_i and v_j, $i, j \in [n]$, are connected by an edge in the random intersection graph $\mathcal{G}(n, m, p)$ if their attribute sets intersect, i.e. $\mathcal{W}(v_i) \cap \mathcal{W}(v_j) \neq \emptyset$. We consider a sequence of random intersection graphs $\mathcal{G}(n, m, p)$, where $m = m_n \to +\infty$ and $p = p_n \to 0$ as $n \to +\infty$.

Note that for each i, $i \in [n]$, the size of $\mathcal{W}(v_i)$ is the binomial random variable $\mathrm{Bin}(m, p)$ with the expected value mp. We may interpret the structure of $\mathcal{G}(n, m, p)$ in another manner. Let $\mathcal{V}(w_i) = \{v \in \mathcal{V} : w_i \in \mathcal{W}(v)\}$ be the set of vertices that chose attribute w_i, $i \in [m]$. Then the edge set of $\mathcal{G}(n, m, p)$ is the union of edges of m independent cliques on vertex sets $\mathcal{V}(w_i)$, $i \in [m]$, with sizes with the binomial distribution $\mathrm{Bin}(n, p)$ and the expected value np.

Now let us discuss some properties of $\mathcal{G}(n, m, p)$ and the chosen range of parameters. In the random intersection graph $\mathcal{G}(n, m, p)$ an edge $v_i v_j$, $i, j \in [n]$, is present with probability $\hat{\mathbf{p}} = 1 - (1-p^2)^m = 1 - e^{-mp^2 + O(mp^4)}$. However edges do not appear independently. In this article we focus on sparse random intersection graphs, i.e. when $\hat{\mathbf{p}} = o(1)$, which is equivalent to $mp^2 = o(1)$. Our focus is motivated by the fact that most interesting results concerning modularity of random graphs concern the sparse case. Moreover we assume that the expected number of edges tends to infinity, i.e. $n^2 mp^2 \to \infty$, as otherwise with positive probability $\mathcal{G}(n, m, p)$ is an empty graph (see for example [14]). In addition, we set $\mathbf{d} = nmp^2 \sim (n-1)\hat{\mathbf{p}}$ which is asymptotically close to the average degree of $\mathcal{G}(n, m, p)$. In some cases we assume that $\mathbf{d} \geq 1$. This is a natural assumption as $\mathbf{d} = 1$ is the phase transition threshold for the emergence of the giant component of $\mathcal{G}(n, m, p)$ for $m \not\asymp n$ (see [2]) and $\mathbf{d} \asymp 1$ is the phase transition threshold in the remaining cases $m \asymp n$ (see [25]). Our interest in the range $\mathbf{d} \geq 1$ is motivated by known results about $G(n, \bar{\mathbf{p}})$, the Erdős–Rényi random graph with independent edges.

The structure of $\mathcal{G}(n, m, p)$ is very diverse and a lot depends on the values np and mp. A long line of research showed that for $m \gg n^3$ the graphs $\mathcal{G}(n, m, p)$ and $G(n, \bar{\mathbf{p}})$ with $\bar{\mathbf{p}}$ close to $1 - e^{-mp^2}$ are equivalent ([11,14,24,35]. We should point out that $\mathcal{G}(n, m, p)$ with high probability is not edgeless for $p = \Omega(1/n\sqrt{m})$ and not a complete graph for $p = O(\sqrt{\ln n/m})$. Therefore $m \gg n^3$ implies mp

very large and np very small. On the other hand, in some range of parameters, when $np \ll 1$, even though $\mathcal{G}(n, m, p)$ and $G(n, \bar{\mathbf{p}})$ are not equivalent, there is some close relation between the models [36,37]. Nevertheless for $np \gg 1$ the models differ a lot (see for example results form [23] and discussion in [14]). One of the motivations for our work was comparison of and discussion about the relation between the results concerning modularity of $\mathcal{G}(n, m, p)$ and $G(n, \bar{\mathbf{p}})$.

Related Work on Modularity. Modularity was defined by Newman and Girvan in 2004 [31,32] and has been extensively studied since.

Definition 1. *Let G be a graph with at least one edge. For a partition $\mathcal{A}$ of $V(G)$ define a modularity score of G as*

$$\mathrm{mod}_{\mathcal{A}}(G) = \sum_{S \in \mathcal{A}} \left(\frac{e(S)}{e(G)} - \left(\frac{\mathrm{vol}(S)}{\mathrm{vol}(G)} \right)^2 \right).$$

Modularity of G is given by

$$\mathrm{mod}(G) = \max_{\mathcal{A}} \mathrm{mod}_{\mathcal{A}}(G),$$

where maximum runs over all the partitions of the set $V(G)$.

It follows straightforward that $\mathrm{mod}(G) \in [0, 1]$. It is considered that "large" modularity, closer to 1, with some exceptions, is related to visible community structure, while modularity close to 0 depicts no community structure. In the context of our research the most important known results concern other random graph models. The modularity of the Erdős–Rényi random graph with independent edges was studied by McDiarmid and Skerman [30]. In particular they showed that for the average degree $n\bar{\mathbf{p}} \leq 1 + o(1)$ the modularity of $G(n, \bar{\mathbf{p}})$ is tending to 1 in probability. It is consistent with other known results concerning modularity of random graphs with no communities and more tree-like structure (see [29]). However we focus on the case above phase transition threshold for the emergence of the giant component, i.e. when $n\bar{\mathbf{p}} \geq C$, for some constant C, and $\bar{\mathbf{p}}$ is bounded away from 1. McDiarmid and Skerman proved that then with high probability $\mathrm{mod}(G(n, \bar{\mathbf{p}})) \asymp 1/\sqrt{n\bar{\mathbf{p}}}$, where $n\bar{\mathbf{p}}$ is asymptotically the average degree of $G(n, \bar{\mathbf{p}})$. This result shows that for average degree $n\bar{\mathbf{p}} \to \infty$ as $n \to \infty$ with high probability $G(n, \bar{\mathbf{p}})$ has modularity $o(1)$. Moreover this result is consistent with modularity $\Theta(1/\sqrt{r})$ for random r–regular graphs for $r \geq 3$ [29] and modularity of the basic preferential attachment model with average degree $2h$ that is $\Omega(1/\sqrt{h})$ [33,34] and $O(\sqrt{\ln h/h})$ [38].

Results. Recall that mp is the average size of the attribute set $\mathcal{W}(v_i)$, $i \in [n]$, of a vertex and np is the average size of the clique $\mathcal{V}(w_i)$, $i \in [m]$, related to an attribute. It is plausible that large modularity (i.e. apparent and visible community structure) should be related to large np and small mp. Since then communities should be defined by attributes and, in average, no vertex (element of the network) is associated with too many such communities. The first of the results concentrates on this case, i.e. $mp \ll 1$ and $np \gg 1$. In fact it follows from the proof that the modularity is in fact closely related to the community structure given by the cliques formed by elements of the network sharing a common attribute.

We recall that assumption $mp^2 = o(1)$ means that the graph is sparse (edge probability tends to 0 as $n \to \infty$) and assumption $n^2 mp^2 \to \infty$ as $n \to \infty$ implies that with high probability $\mathcal{G}(n, m, p)$ has at least one edge. As natural, these assumptions appear in all the theorems.

Theorem 1. *Let $mp^2 = o(1)$ and $n^2 mp^2 \to \infty$ as $n \to \infty$. There exists $C > 0$ such that for all $\varepsilon > 0$ there exists A_ε such that if*

$$npe^{-mp} \geq A_\varepsilon,$$

then with probability at least $1 - \varepsilon$

$$\mathrm{mod}(\mathcal{G}(n, m, p)) \geq (1 - C\varepsilon)e^{-mp}.$$

Remark 1. In the proof $A_\varepsilon = 3\varepsilon^{-2}\ln(4/\varepsilon^2)$ and $C \leq 13$. However we made no attempt to make the constant optimal.

Corollary 1. *Let $mp^2 = o(1)$, $n^2 mp^2 \to \infty$, and $np \to \infty$ as $n \to \infty$. Then with high probability*

$$\mathrm{mod}(\mathcal{G}(n, m, p)) \geq (1 + o(1))e^{-mp}.$$

In particular, for $mp = o(1)$

$$\mathrm{mod}(\mathcal{G}(n, m, p)) = 1 + o(1).$$

However, the above corollary does not provide any relevant information in the case when $mp \to \infty$. In fact, the following theorem shows that when not only $np \to \infty$ but also $mp \to \infty$ as $n \to \infty$, it is possible for $\mathcal{G}(n, m, p)$ to not have high modularity.

Theorem 2. *Let $mp^2 = o(1)$, $n^2 mp^2 \to \infty$, $n = o(m)$, and $np/\ln \frac{m}{n} \to \infty$ as $n \to \infty$. Then with high probability*

$$\mathrm{mod}(\mathcal{G}(n, m, p)) = O\left(\left(\frac{\ln(m/n)}{np}\right)^{1/2} + \frac{n}{m} + \omega mp^2\right),$$

for any $\omega \to \infty$ arbitrarily slowly as $n \to \infty$.

We should stress that in the above considered case not only $np \to \infty$, but also $mp \to \infty$, as $n \to \infty$.

The following two results are an attempt to relate the known results concerning $G(n, \bar{\mathbf{p}})$ to modularity of $\mathcal{G}(n, m, p)$. Recall that for $G(n, \bar{\mathbf{p}})$ with the average degree $n\bar{\mathbf{p}} = \Omega(1)$ and $\bar{\mathbf{p}} = o(1)$ modularity is asymptotically $\Theta(1/\sqrt{n\bar{\mathbf{p}}})$.

Theorem 3. *Let* $np = o(1)$, $mp^2 = o(1)$, *and* $\mathbf{d} = nmp^2 \geq 1$. *Then for any* $A > 0$ *there exists a constant* C *such that for any* $\omega = \omega_n \to \infty$ *as* $n \to \infty$ *with high probability*

$$\mathrm{mod}(\mathcal{G}(n,m,p)) \leq \frac{C}{\sqrt{\mathbf{d}}} + O\left((np)^A + \omega m p^2\right).$$

Corollary 2. *Let* $m \geq n^2$, $mp^2 = o(n^{-1/3})$, *and* $\mathbf{d} = nmp^2 \geq 1$. *Then there exists a constant* C *such that*

$$\mathrm{mod}(\mathcal{G}(n,m,p)) \leq \frac{C}{\sqrt{\mathbf{d}}}.$$

The last theorem relates both models $\mathcal{G}(n,m,p)$ and $G(n,\bar{\mathbf{p}})$ more directly.

Theorem 4. *Let* $np = o(1)$ *and*

$$\bar{\mathbf{p}} = 1 - e^{-m\hat{q}/\binom{n}{2}}, \text{ where } \hat{q} = 1 - (1-p)^n - np(1-p)^{n-1}.$$

Then with high probability for any $\delta > 0$

$$\mathrm{mod}(\mathcal{G}(n,m,p)) = (1 + o(1))\mathrm{mod}(G(n,\bar{\mathbf{p}})) + O((np)^{1-\delta}).$$

The remaining part of the article is organised as follows. In the next sections we give auxiliary results. In particular, Sect. 2 states one known and one new useful result about modularity. In Sect. 3 we give some general results about the edge counts in $\mathcal{G}(n,m,p)$. Section 4 states known probabilistic inequalities that are used in the proofs. Due to limited space, in the Sects. 5 and 6 we give the proofs of only two theorems. We finish with concluding remarks.

2 Modularity Lemmas

We will use a result on modularity by Dinh and Thai.

Lemma 1. ([13], **Lemma 1**). *Let* G *be a graph with at least one edge and let* $k \in \mathbb{N} \setminus \{1\}$. *Then*

$$\mathrm{mod}(G) \leq \frac{k}{k-1} \max_{\mathcal{A}:|\mathcal{A}|\leq k} \mathrm{mod}_{\mathcal{A}}(G).$$

We prove the following corollary of this lemma.

Corollary 3. *Let* G *be a graph with at least one edge then*

$$\mathrm{mod}(G) \leq 4 \max_{S \subseteq \mathcal{V}, |S| \geq n/2} \left(\frac{e(S)}{e(G)} - \left(\frac{\mathrm{vol}(S)}{\mathrm{vol}(G)}\right)^2 \right).$$

Proof. First note that

$$e(S) = e(\bar{S}) + e(G) - \text{vol}(\bar{S}),$$
$$\text{vol}^2(S) = \big(\text{vol}(G) - \text{vol}(\bar{S})\big)^2 = \text{vol}(G)\big(\text{vol}(G) - 2\text{vol}(\bar{S})\big) + \text{vol}^2(\bar{S}).$$

Therefore

$$\frac{e(S)}{e(G)} - \left(\frac{\text{vol}(S)}{\text{vol}(G)}\right)^2 = \frac{e(\bar{S})}{e(G)} - \left(\frac{\text{vol}(\bar{S})}{\text{vol}(G)}\right)^2. \tag{1}$$

Note that for $|\mathcal{A}| = 1$ we have $\text{mod}_{\mathcal{A}}(G) = 0$. Thus, as either $S \geq n/2$ or $\bar{S} > n/2$, by Definition 1, Lemma 1, and (1)

$$\text{mod}(G) \leq 2 \max_{\mathcal{A}, |\mathcal{A}|=2} \text{mod}_{\mathcal{A}}(G)$$

$$= 2 \max_{S \subseteq \mathcal{V}} \left(\frac{e(S)}{e(G)} - \left(\frac{\text{vol}(S)}{\text{vol}(G)}\right)^2 + \frac{e(\bar{S})}{e(G)} - \left(\frac{\text{vol}(\bar{S})}{\text{vol}(G)}\right)^2\right)$$

$$= 2 \max_{S \subseteq \mathcal{V}, |S| \geq n/2} 2\left(\frac{e(S)}{e(G)} - \left(\frac{\text{vol}(S)}{\text{vol}(G)}\right)^2\right).$$

3 Counting Edges in $\mathcal{G}(n,m,p)$

Recall that, as mentioned in the introduction, in $\mathcal{G}(n,m,p)$ each edge appears with probability

$$1 - (1-p^2)^m = (1 - e^{-(1+O(p^2))mp^2}) = mp^2(1 + O(mp^2)), \text{ for } mp^2 = o(1).$$

Therefore

$$\mathbb{E}e(\mathcal{G}(n,m,p)) = \binom{n}{2} mp^2(1 + O(mp^2)) = \frac{1}{2} n^2 mp^2(1 + O(mp^2 + n^{-1})).$$

Moreover, for $mp^2 = o(1)$,

$$\text{Var}(e(\mathcal{G}(n,m,p))) \sim \binom{n}{2} mp^2 + (n)_3 mp^3 \asymp n^2 mp^2(1 + np).$$

By Chebyshev's inequality we get that for $mp^2 \ll 1$ and $n^2 mp^2 \gg 1$ with high probability

$$\text{vol}(\mathcal{G}(n,m,p)) = 2e(\mathcal{G}(n,m,p))$$
$$= n^2 mp^2 \left(1 + O\left(n^{-1} + mp^2 + \omega(1+np)(n^2 mp^2)^{-1/2}\right)\right) \sim n^2 mp^2. \tag{2}$$

Recall that $\omega = \omega_n$ is a function tending to infinity arbitrarily slowly as $n \to \infty$. For $S \subseteq \mathcal{V}$, let

$$X_{i,S} = |\mathcal{V}(w_i) \cap S|, \quad V_i = X_{i,\mathcal{V}} = |\mathcal{V}(w_i)|, \quad i \in [m].$$

By definition, each attribute $w_i \in \mathcal{W}$, $i \in [m]$, generates a clique of size V_i in $\mathcal{G}(n, m, p)$. Moreover each edge of $\mathcal{G}(n, m, p)$ is included in at least one clique. However some edges might be included in two or more such cliques. Let us denote by $E_1 = |\{vv' \in E(\mathcal{G}(n, m, p)) : \exists_{i,j \in [m]} v, v' \in \mathcal{V}(w_i) \cap \mathcal{V}(w_j), i \neq j\}|$ the number of such edges that contribute to more than one clique. Then $\mathbb{E}E_1 \leq n^2 m^2 p^4$ and by Markov's inequality and (2) we get that, for $mp^2 \ll 1$ and $n^2 m p^2 \gg 1$, with high probability

$$E_1 = O\left(\omega n^2 m^2 p^4\right) = O(\omega m p^2) e(\mathcal{G}(n, m, p)). \tag{3}$$

Now we may relate the number of edges $e(S)$, $e(S, \bar{S})$, and the volume $\text{vol}(S)$ in $\mathcal{G}(n, m, p)$ to $X_{i,S}$ and $X_{i,\bar{S}}$. Recall that $X_{i,S}$ is the number of vertices in S that chose attribute w_i. Therefore $\binom{X_{i,S}}{2}$ is the number of edges included in S that are in the clique of vertices that chose w_i. Each edge from S is included in at least one such clique. Similarly $X_{i,S} X_{i,\bar{S}}$ is the number of edges with one end in S and the other in $\bar{S}$ and included in the clique related to w_i. Therefore, taking into account that some edges might be included in two cliques (see the above discussion about E_1), we get

$$e(S) \leq \sum_{i \in [m]} \binom{X_{i,S}}{2} = \frac{1}{2} \sum_{i \in [m]} (X_{i,S})_2, \tag{4}$$

$$e(S, \bar{S}) \geq \sum_{i \in [m]} X_{i,S} X_{i,\bar{S}} - E_1, \text{ and} \tag{5}$$

$$\text{vol}(S) = 2e(S) + e(S, \bar{S}) \geq \sum_{i \in [m]} X_{i,S}(V_i - 1) - 2E_1. \tag{6}$$

4 Auxiliary Lemmas

In this section we give some classical inequalities that are used in the remaining part of the article. We start with Freedman's inequality. We state it here in the from that is convenient for our purposes (see corollary of Lemma 22 of [42] by Warnke or [3] by Bennett and Dudek).

Lemma 2. Let $M_0, M_1, \ldots, M_m$ be a martingale with respect to a filtration $\mathcal{F}_0 \subseteq \mathcal{F}_1 \subseteq \ldots \subseteq \mathcal{F}_n$. Let $|M_i - M_{i-1}| \leq C$, for all $i \in [m]$, and $V = \sum_{i=1}^m \text{Var}(M_i - M_{i-1}|\mathcal{F}_{i-1})$. Then

$$\Pr(M_m \geq M_0 + \lambda) \leq \exp\left\{-\frac{\lambda^2}{2V(1 + C\lambda/(3V))}\right\}.$$

We will also use Chernoff's bound for the binomial distribution (see for example Theorem 2.1 and Corollary 2.3 in [18])

Lemma 3. *Let X has the binomial distribution $\mathrm{Bin}\,(n,p)$, then for $t > 0$*

$$\Pr(X \geq np + t) \leq \exp\left(-\frac{t^2}{3np}\right), \quad \textit{for } t/np \leq 3/2;$$

$$\Pr(X \leq np - t) \leq \exp\left(-\frac{t^2}{2np}\right).$$

5 Proof of Theorem 1

Recall that $mp^2 \ll 1$ and $n^2 mp^2 \gg 1$. Denote by

$$\tilde{\mathcal{V}}_i = \{v \in \mathcal{V}(w_i) : |\mathcal{W}(v)| = 1\}, \quad \tilde{V}_i = |\tilde{\mathcal{V}}_i|, i \in [m]$$

the sets of vertices that chose exactly one attribute w_i, $i \in [m]$, and their sizes. Note that $\tilde{V}_i$, $i \in [m]$, has the binomial distribution $\mathrm{Bin}\,(n, p(1-p)^{m-1})$, i.e. $\mathbb{E}(\tilde{V}_i) \sim npe^{-mp}$, $i \in [m]$. Set $\varepsilon \in (0,1)$ and $A_\varepsilon = 3\varepsilon^{-2} \ln(4/\varepsilon^2)$. Let np and mp be such that $npe^{-mp} > A_\varepsilon$ (i.e. also $np > A_\varepsilon$). As V_i and $\tilde{V}_i$, $i \in [m]$, have binomial distributions, by Chernoff's inequality (Lemma 3), for all $i \in [m]$,

$$\Pr\left(\left|\tilde{V}_i - npe^{-mp}\right| \geq \varepsilon npe^{-mp}\right) \leq 2\exp\left(-\frac{\varepsilon^2 A_\varepsilon}{3}\right) \leq \frac{\varepsilon^2}{2} \text{ and}$$

$$\Pr\left(|V_i - np| \geq \varepsilon np\right) \leq 2\exp\left(-\frac{\varepsilon^2 A_\varepsilon}{3}\right) \leq \frac{\varepsilon^2}{2}.$$

Let $\mathcal{M}_1 = \left\{i \in [m] : |\tilde{V}_i - npe^{-mp}| \leq \varepsilon npe^{-mp} \text{ and } |V_i - np| \leq \varepsilon np\right\}$. Then by the union bound and the above concentration results

$$\mathbb{E}|[m] \setminus \mathcal{M}_1|$$
$$\leq \sum_{i \in [m]} \Pr\left(|\tilde{V}_i - npe^{-mp}| \geq \varepsilon npe^{-mp} \text{ or } |V_i - np| \geq \varepsilon np\right) \leq \varepsilon^2 m. \quad (7)$$

Therefore by Markov's inequality $\Pr\left(|[m] \setminus \mathcal{M}_1| \geq \varepsilon m\right) \leq \varepsilon$, i.e.

$$|\mathcal{M}_1| \geq (1-\varepsilon)m \text{ with probability at least } 1-\varepsilon. \quad (8)$$

Let us define a partition of the vertex set

$$\mathcal{A} = \{\tilde{\mathcal{V}}_i : i \in \mathcal{M}_1\} \cup \{\bar{\mathcal{V}}\}, \text{ where } \bar{\mathcal{V}} = \mathcal{V} \setminus \bigcup_{i \in \mathcal{M}_1} \tilde{\mathcal{V}}_i.$$

Now set G, an instance of the random intersection graph $\mathcal{G}\,(n,m,p)$ with properties

$$|\mathcal{M}_1| \geq (1-\varepsilon)m \quad \text{and} \quad |2e(G) - n^2 mp^2| \leq \varepsilon n^2 mp^2. \quad (9)$$

Note that by (2) and (8), for any $\varepsilon \in (0,1)$, $\mathcal{G}\,(n,m,p)$ has the above stated properties with probability at least $1 - 2\varepsilon$.

First let us consider $S = \tilde{\mathcal{V}}_i$, $i \in \mathcal{M}_1$. Note that by the definition of $\mathcal{M}_1$

$$2e(S) = 2e(\tilde{\mathcal{V}}_i) = \tilde{V}_i(\tilde{V}_i - 1) \geq (1-\varepsilon)^2 (npe^{-mp})^2 \left(1 - \frac{1}{(1-\varepsilon)npe^{-mp}}\right)$$

$$\geq (1-\varepsilon)^3 (np)^2 e^{-2mp} \geq (1-\varepsilon)^4 \frac{2e(G)}{m} e^{-2mp}. \quad (10)$$

The second last inequality follows as $(1-\varepsilon)npe^{-mp} > (1-\varepsilon)A_\varepsilon > \varepsilon^{-1}$ for all $\varepsilon \in (0, 0.8)$. Moreover

$$\mathrm{vol}(S) = \mathrm{vol}(\tilde{\mathcal{V}}_i) = \tilde{V}_i(V_i - 1) \leq (1+\varepsilon)^2 npe^{-mp} np \leq (1+\varepsilon)^3 \frac{\mathrm{vol}(G)}{m} e^{-mp}.$$

Therefore, for $S = \tilde{\mathcal{V}}_i$, $i \in \mathcal{M}_1$, and large enough m we get

$$\frac{e(\tilde{\mathcal{V}}_i)}{e(G)} - \left(\frac{\mathrm{vol}(\tilde{\mathcal{V}}_i)}{\mathrm{vol}(G)}\right)^2 \geq (1-5\varepsilon)\frac{e^{-2mp}}{m}. \quad (11)$$

For $S = \bar{\mathcal{V}}$, using calculations for $\tilde{\mathcal{V}}_i$, $i \in \mathcal{M}_i$,

$$e(\bar{S}) = \sum_{i \in \mathcal{M}_1} e(\tilde{\mathcal{V}}_i) \geq (1-\varepsilon)m(1-\varepsilon)^4 \frac{e(G)e^{-2mp}}{m} \geq (1-5\varepsilon)e(G)e^{-2mp},$$

$$(\mathrm{vol}(\bar{S}))^2 = \left(\sum_{i \in \mathcal{M}_1} \mathrm{vol}(\tilde{\mathcal{V}}_i)\right)^2 \leq (1+\varepsilon)^6 \mathrm{vol}^2(G) e^{-2mp}.$$

Combining the above results with (1) we get for $\varepsilon < 1/2$ and large enough m

$$\frac{e(\bar{\mathcal{V}})}{e(G)} - \left(\frac{\mathrm{vol}(\bar{\mathcal{V}})}{\mathrm{vol}(G)}\right)^2 \geq -26\varepsilon e^{-2mp}. \quad (12)$$

Recall that $\mathcal{G}(n,m,p)$ with probability at least $1 - 2\varepsilon$ has properties (9). As G with properties (9) fulfil (11) and (12), we get that with probability at least $1 - 2\varepsilon$

$$\mathrm{mod}(\mathcal{G}(n,m,p)) \geq \mathrm{mod}_\mathcal{A}(\mathcal{G}(n,m,p))$$

$$\geq \sum_{i \in \mathcal{M}_i} (1-5\varepsilon)\frac{e^{-2mp}}{m} - 26\varepsilon e^{-2mp} \geq (1-31\varepsilon) e^{-2mp}.$$

6 Proof of Theorem 3

Recall that we assume that $np = o(1)$, $mp^2 = o(1)$, and $\mathbf{d} = nmp^2 \geq 1$. Let us set S such that $|S| = s \geq n/2$. Recall that $V_i = |\mathcal{V}(w_i)|$ and $X_i = X_{i,S}$ have the binomial distributions $\mathrm{Bin}(n,p)$ and $\mathrm{Bin}(s,p)$, resp.. Moreover define $Y_i = X_{i,\bar{S}} \sim \mathrm{Bin}(n-s,p)$, $i \in [m]$. By definition $X_1, \ldots, X_m, Y_1, \ldots, Y_m$ are all independent random variables.

We set a constant $A \geq 6$ and define auxiliary random variables
$$\tilde{X}_i = X_i \mathbb{I}_{X_i \leq A} \quad \text{and} \quad \tilde{Y}_i = Y_i \mathbb{I}_{Y_i \leq A}, \text{ for } i \in [m].$$
For $l = 1, 2, 3, 4$ and $i \in [m]$

$$\mathbb{E}(X_i)_l \mathbb{I}_{X_i > A} \leq \sum_{k=A}^{s} k^l \Pr(V_i \geq k) \leq \sum_{k=A}^{s} k^l \binom{s}{k} p^k \leq \sum_{k=A}^{s} k^l \left(\frac{esp}{k}\right)^k$$

$$\leq \sum_{k=A}^{s} \frac{(esp)^k}{k^{k-l}} \leq A^l \left(\frac{esp}{A}\right)^A \frac{1}{1-esp} = (1+O(sp)) A^l \left(\frac{esp}{A}\right)^A = O((sp)^A) \quad (13)$$

and similarly $\mathbb{E}(Y_i)_l \mathbb{I}_{Y_i > A} = O(((n-s)p)^A)$. Let

$$E_2 = \sum_{i \in [m]} (V_i)_2 \mathbb{I}_{V_i > A} \quad (14)$$

be the upper bound on number of edges contained in cliques of size larger than A. Substituting $s = n$, i.e. $S = \mathcal{V}$, to (13) we get

$$\mathbb{E}(E_2) = \mathbb{E}\left(\sum_{i \in [m]} (V_i)_2 \mathbb{I}_{V_i > A}\right) = O\left(m(np)^A\right).$$

Therefore by Markov's inequality, for $B < A$

$$\Pr\left(E_2 \geq (np)^B m\right) \leq (np)^{A-B} = o(1). \quad (15)$$

It is well known that $\mathbb{E}(X_i)_l = (s)_l p^l$ and $\mathbb{E}(Y_i)_l = (n-s)_l p^l$, $l = 1, 2, 3, 4$, $i \in [m]$. The observations made so far and standard calculations lead us to the following conclusions that for $A \geq 6$.

$$\mathbb{E}((\tilde{X}_i)_2) = s^2 p^2 - sp^2 + O((np)^A), \quad \mathbb{E}(\tilde{X}_i \tilde{Y}_i) = s(n-s)p^2 + O((np)^A),$$
$$\mathrm{Var}((\tilde{X}_i)_2) = 2s^2 p^2 (1 + O(np + n^{-1})), \quad \mathrm{Var}(\tilde{X}_i \tilde{Y}_i) = s(n-s)p^2(1 + O(np)). \quad (16)$$

We define two martingales, with the filtration $(\mathcal{F}_t)_{t \in [m]}$ defined as follows $\mathcal{F}_t = \sigma(\{X_i, Y_i : i \in [t]\})$

$$M_0 = M_0(S) = 0, \quad M_t = M_t(S) = \sum_{i=1}^{t}((\tilde{X}_i)_2 - \mathbb{E}(\tilde{X}_i)_2),$$

$$\bar{M}_0 = \bar{M}_0(S) = 0, \quad \bar{M}_t = \bar{M}_t(S) = \sum_{i=1}^{t}(\tilde{X}_i \tilde{Y}_i - \mathbb{E}(\tilde{X}_i \tilde{Y}_i)),$$

for $t = 1, 2, \ldots, m$. Then, by (4), (5), (15), with high probability

$$2e(S) \leq \sum_{i=1}^{m}(\tilde{X}_i)_2 + E_2 = M_m + s^2 m p^2 + O((np)^B),$$

$$\mathrm{vol}(S) = 2e(S) + e(S, \bar{S}) \geq \sum_{i=1}^{m}(\tilde{X}_i)_2 + \sum_{i=1}^{t} \tilde{X}_i \tilde{Y}_i - 2E_1 \quad (17)$$
$$= M_m + \bar{M}_m + snmp^2 + O((np)^B) - 2E_1.$$

Define $\lambda_S = \varepsilon_n snmp^2 = \varepsilon_n s\mathbf{d}$ and $\varepsilon_n = \frac{3A^3}{\sqrt{\mathbf{d}}}$, where $\mathbf{d} = nmp^2$. We apply Lemma 2 to both martingales. For M_t by the definition of $\tilde{X}_i$ and (16)

$$|M_i - M_{i-1}| = |(\tilde{X}_i)_2 - \mathbb{E}(\tilde{X}_i)_2| \leq A^2;$$
$$\mathrm{Var}(M_i - M_{i-1}|\mathcal{F}_{i-1}) = \mathrm{Var}((\tilde{X}_i)_2) \sim 2\,s^2p^2, \text{i.e.} \quad V \sim 2\,s^2mp^2.$$

Then

$$\frac{C\lambda_S}{3V} \leq \frac{A^2 \varepsilon_n nsmp^2}{(1+o(1))6\,s^2mp^2} \leq \frac{1}{2}A^2\varepsilon_n, \text{ for } s \geq n/2.$$

Then by Lemma 2, for $\mathbf{d} \geq 1$, $A \geq 6$, and $s \geq n/2$, we have

$$\Pr(M_m \geq \lambda_S) = \Pr(M_m \geq M_0 + \lambda_S)$$
$$\leq \exp\left(-\frac{\varepsilon_n^2 s^2n^2 m^2p^4}{4(1+o(1))s^2mp^2(1+A^2\varepsilon_n/2)}\right)$$
$$\leq \exp\left(-\frac{9A^6}{5(1+3A^5/2)}n\right) \leq \exp(-0.79n) = o(2^{-n}).$$

Moreover, as $(-M_t)_{t \in [m]}$ is also a martingale, in an analogous manner we get

$$\Pr(M_m \leq -\lambda_S) = \Pr(-M_m \geq -M_0 + \lambda_S) = o(2^{-n}) \quad \text{and similarly}$$
$$\Pr(\bar{M}_m \leq -\lambda_S) \leq \exp\left(-\frac{3A^6}{(1+A^2/2)}\right) = o(2^{-n}).$$

Therefore $\Pr\left(\exists_{S, |S| \geq n/2} |M_S| \geq \lambda_S \text{ or } \bar{M}_S \leq -\lambda_S\right) \leq 2^n \cdot o(2^{-n}) = o(1)$ and, as a consequence, with high probability

$$|M_S| \leq \lambda_S \text{ and } \bar{M}_S \geq -\lambda_S. \tag{18}$$

Let us take an instance G of $\mathcal{G}(n, m, p)$ with properties (18) and (2). With high probability $\mathcal{G}(n, m, p)$ has those properties. Set $\lambda = \lambda_\mathcal{V} = \varepsilon_n n^2 mp^2 = \varepsilon_n n\mathbf{d} \geq \lambda_S$. By (17) we have for G and all $S \subseteq \mathcal{V}$

$$2e(S) \leq \lambda + s^2mp^2 + O(\mathrm{err}) \quad \text{and} \quad \mathrm{vol}(S) \geq -2\lambda + snmp^2 - O(\mathrm{err}),$$

where we use a shorthand notation $\mathrm{err} := (np)^B + \omega n^2 m^2 p^4$.

By definition $\lambda \asymp n\sqrt{\mathbf{d}} = n^{3/2}m^{1/2}p$ and $1 \leq \mathbf{d} = nmp^2 = o(n)$. Then, for ω tending slowly to infinity

$$nmp^2 = \mathbf{d} = o(\lambda), \quad \omega nm^{1/2}p = o(\lambda), \quad \text{and} \quad n^2m^2p^4 \leq \mathrm{err}.$$

Therefore by (2) we have

$$2e(G) = n^2mp^2 + o(\lambda) + O(\mathrm{err}). \tag{19}$$

We recall that $\lambda = n^{3/2}m^{1/2}p = n^2mp^2/\sqrt{\mathbf{d}} \leq n^2mp^2$. Moreover $\mathrm{err} = o(n^2mp^2)$. Before we substitute the above estimates to the formula related to the modularity, we perform some calculations

$$2e(S) \cdot 2e(G) \leq (s^2mp^2 + \lambda + O(\text{err}))(n^2mp^2 + o(\lambda) + O(\text{err}))$$
$$= s^2n^2m^2p^4 + \lambda(1+o(1))n^2mp^2 + n^2mp^2 O(\text{err}),$$
$$(\text{vol } S)^2 \geq (snmp^2 - \lambda - O(\text{err}))^2$$
$$\geq s^2n^2m^2p^4 - 2\lambda n^2mp^2 - n^2mp^2 O(\text{err}).$$

Thus we get for all $S \subseteq \mathcal{V}$

$$\frac{e(S)}{e(G)} - \frac{(\text{vol } S)^2}{(\text{vol } G)^2} = \frac{2e(S) \cdot 2e(G) - (\text{vol } S)^2}{4e(G)}$$
$$= (1+o(1))\frac{9A^3}{\sqrt{\mathbf{d}}} + O\left((np)^{B-2} + \omega mp^3\right).$$

This combined with Corollary 3 gives the statement of Theorem 3.

7 Concluding Remarks

We studied modularity of the classical random intersection graph introduced in [23]. We showed that, when each vertex of the network does not have too many attributes and each attribute is possessed by many vertices, then there is an apparent community structure in the network detected by the modularity. We have also given an example in which there are still large communities related to common attributes, however the number of attributes of each vertex is large. In this case the community structure is no longer detected by the modularity, i.e. the modularity tends to 0 as $n \to \infty$. It would be interesting to understand the behaviour of the random intersection graphs in the case when both np and mp tends to infinity, but $m = o(n)$. This case is not covered by the above mentioned results. We also studied a relation between the modularity of $\mathcal{G}(n,m,p)$ and classical Erdős–Rény random graph with independent edges.

We think that these results give an insight into the relation between community structure of networks based on a bipartite graph of an affiliation network. We studied only the most classical random intersection graph model $\mathcal{G}(n,m,p)$. We hope that the research will follow and more interesting models, form the point of view of applications, will be considered. Here we have in mind such models that not only have large clustering but also have power law degree distribution and possibly more properties of complex networks.

References

1. Bagrow, J.P.: Communities and bottlenecks: trees and treelike networks have high modularity. Phys. Rev. E **85**, 066118 (2012)
2. Behrisch, M.: Component evolution in random intersection graphs. Electron. J. Comb. **14**(1), R17 (2007)

3. Bennett, P., Dudek, A.: A gentle introduction to the differential equation method and dynamic concentration. Discret. Math. **345**(12), 113071 (2022)
4. Blondel, V.D., Guillaume, J.L., Lambiotte, R., Lefebvre, E.: Fast unfolding of communities in large networks. J. Stat. Mech: Theory Exp. **2008**(10), P10008 (2008)
5. Bloznelis, M., Godehardt, E., Jaworski, J., Kurauskas, V., Rybarczyk, K.: Recent Progress in Complex Network Analysis: Models of Random Intersection Graphs. In: Data Science, Learning by Latent Structures, and Knowledge Discovery, pp. 69–78. Springer, Heidelberg (2015)
6. Bloznelis, M., Godehardt, E., Jaworski, J., Kurauskas, V., Rybarczyk, K.: Recent Progress in Complex Network Analysis: Properties of Random Intersection Graphs. In: Data Science, Learning by Latent Structures, and Knowledge Discovery, pp. 79–88. Springer, Heidelberg (2015)
7. Bloznelis, M., Karjalainen, J., Leskelä, L.: Assortativity and bidegree distributions on Bernoulli random graph superpositions. Probab. Eng. Inf. Sci. **36**(4), 1188–1213 (2021)
8. Bloznelis, M., Karoński, M.: Random intersection graph process. Internet Math. **11**(4–5), 385–402 (2014)
9. Bloznelis, M., Leskelä, L.: Clustering and percolation on superpositions of Bernoulli random graphs. Random Struct. Algorithms **63**(2), 283–342 (2023)
10. Brandes, U., et al.: On modularity clustering. IEEE Trans. Knowl. Data Eng. **20**(2), 172–188 (2008)
11. Brennan, M., Bresler, G., Nagaraj, D.: Phase transitions for detecting latent geometry in random graphs. Probab. Theory Relat. Fields 1215–1289 (2020)
12. Chellig, J., Fountoulakis, N., Skerman, F.: The modularity of random graphs on the hyperbolic plane. J. Complex Netw. **10**(1), cnab051 (2021)
13. Dinh, T.N., Thai, M.T.: Finding community structure with performance guarantees in scale-free networks. In: 2011 IEEE Third International Conference on Privacy, Security, Risk and Trust and 2011 IEEE Third International Conference on Social Computing, pp. 888–891 (2011)
14. Fill, J.A., Scheinerman, E.R., Singer-Cohen, K.B.: Random intersection graphs when $m = \omega(n)$: an equivalence theorem relating the evolution of the G(n, m, p) and G(n, p) models. Random Struct. Algorithms **16**, 156–176 (2000)
15. Gilad, G., Sharan, R.: From Leiden to Tel-Aviv University (TAU): exploring clustering solutions via a genetic algorithm. PNAS Nexus **2**(6), pgad180 (2023)
16. Godehardt, E., Jaworski, J.: Two models of random intersection graphs for classification. In: Opitz, O., Schwaiger, M. (eds.) Studies in Classifcation, Data Analysis and Knowledge Organization, vol. 22, pp. 67–81. Springer (2003)
17. Guillaume, J.-L., Latapy, M.: Bipartite structure of all complex networks. Inf. Process. Lett. **90**(5), 215–221 (2004)
18. Janson, S., Łuczak, T., Ruciński, A.: Random Graphs. Wiley, Hoboken (2000)
19. Kamiński, B., Misiorek, P., Prałat, P., Théberge, F.: Modularity based community detection in hypergraphs. J. Complex Netw. **12**(5), 26 (2024). Paper No. cnae041
20. Kamiński, B., Pankratz, B., Prałat, P., Théberge, F.: Modularity of the abcd random graph model with community structure. J. Complex Netw. **10**(6) (2022)
21. Kamiński, B., Poulin, V., Prałat, P., Szufel, P., Théberge, F.: Clustering via hypergraph modularity. PLoS ONE **14**, e0224307 (2019)
22. Kamiński, B., Prałat, P., Théberge, F.: Mining Complex Networks. Chapman and Hall/CRC (2021)
23. Karoński, M., Scheinerman, E.R., Singer-Cohen, K.B.: On random intersection graphs: the subgraph problem. Comb. Probab. Comput. **8**, 131–159 (1999)

24. Kim, J.H., Lee, S.J., Na, J.: On the total variation distance between the binomial random graph and the random intersection graph. Random Struct. Algorithms **52**(4), 662–679 (2018)
25. Lagerås, A.N., Lindholm, M.: A note on the component structure in random intersection graphs with tunable clustering. Electron. J. Comb. **15**(1), N10 (2008)
26. Lasoń, M., Sulkowska, M.: Modularity of minor-free graphs. J. Graph Theory **102**(4), 728–736 (2023)
27. Majstorović, S., Stevanović, D.: A note on graphs whose largest eigenvalues of the modularity matrix equals zero. Electron. J. Linear Algebra **27** (2014)
28. McDiarmid, C., Skerman, F.: Modularity in random regular graphs and lattices. Electron. Notes Discrete Math. **43**, 431–437 (2013)
29. McDiarmid, C., Skerman, F.: Modularity of regular and treelike graphs. J. Complex Netw. **6**(4), 596–619 (2018)
30. McDiarmid, C., Skerman, F.: Modularity of Erdős-Rényi random graphs. Random Struct. Algorithms **57**(1), 211–243 (2020)
31. Newman, M.E.J.: Networks: An Introduction. Oxford University Press, Oxford (2010)
32. Newman, M., Girvan, M.: Finding and evaluating community structure in networks. Phys. Rev. E **69**(2), 026113 (2004)
33. Ostroumova Prokhorenkova, L., Prałat, P., Raigorodskii, A.: Modularity of complex networks models. In: Bonato, A., Graham, F.C., Prałat, P. (eds.) WAW 2016. LNCS, vol. 10088, pp. 115–126. Springer, Cham (2016)
34. Prokhorenkova, L., Prałat, P., Raigorodskii, A.M.: Modularity in several random graph models. Electron. Notes Discrete Math. **61**, 947–953 (2017)
35. Rybarczyk, K.: Equivalence of the random intersection graph and G(n, p). Random Struct. Algorithms **38**, 205–234 (2011)
36. Rybarczyk, K.: Sharp threshold functions for random intersection graphs via a coupling method. Electron. J. Comb. **18**(1), P36 (2011)
37. Rybarczyk, K.: The coupling method for inhomogeneous random intersection graphs. Electron. J. Combin. **24**(2), P2.10 (2017)
38. Rybarczyk, K., Sulkowska, M.: Modularity of preferential attachment graphs. Preprint (2025). arxiv.org/abs/2501.06771
39. Spirakis, P.G., Nikoletseas, S., Raptopoulos, C.: A guided tour in random intersection graphs. In: Fomin, F.V., Freivalds, R., Kwiatkowska, M., Peleg, D. (eds.) ICALP 2013. LNCS, vol. 7966, pp. 29–35. Springer, Heidelberg (2013)
40. Traag, V.A., Waltman, L., van Eck, N.J.: From Louvain to Leiden: guaranteeing well-connected communities. Sci. Rep. **9**(5233) (2019)
41. van der Hofstad, R., Komjáthy, J., Vadon, V.: Phase transition in random intersection graphs with communities. Random Struct. Algorithms **60**(3), 406–461 (2021)
42. Warnke, L.: On the method of typical bounded differences. Combin. Probab. Comput. **25**, 269–299 (2022)

The Size of the Giant in Inhomogeneous Random Graphs of Preferential Attachment Type

Peter Mörters and Lucas Schätze[✉]

Mathematisches Institut, Universität zu Köln, Weyertal 86-90, 50931 Köln, Germany
{p.moerters,lschaet2}@uni-koeln.de

Abstract. For the inhomogeneous random graph with kernel of preferential attachment type and degree distribution with power-law exponent $\tau \in (2,3)$ we study the decay of the size of the giant component when the edge density approaches zero. It turns out that the giant component is significantly smaller than for the inhomogeneous random graph with a kernel of rank one.

Keywords: Scale-free network · preferential attachment · small giant · phase transition

1 Introduction

The inhomogeneous random graph with kernel of preferential attachment type is a solvable model, which shares many features of more involved preferential attachment models. If set-up with a power-law exponent $\tau \in (2,3)$ the model is *robust* in the sense that there is a connected component comprising an asymptotically positive proportion of the vertices, no matter how small the edge density. However, the asymptotic proportion of vertices in this giant component decreases very quickly to zero when the edge density decreases to zero. Our result identifies the exact speed at which this happens.

Inhomogeneous random graphs cf. [3], [9, Chapter 3.2], [14], are parametrised by a symmetric, continuous kernel

$$\kappa \colon (0,1] \times (0,1] \to [0,\infty).$$

Given the kernel, the graph $\mathcal{G}_n$ has vertex set $\{1,\ldots,n\}$ and we sometimes refer to the number identifying a vertex as its index. Any pair of distinct vertices $i,j \in \{1,\ldots,n\}$ is connected by an edge, independently with probability

$$p_{i,j} := \frac{1}{n}\kappa\Big(\frac{i}{n},\frac{j}{n}\Big) \wedge 1.$$

In most preferential attachment models the probability at which a vertex arriving at time j connects to an earlier vertex i is proportional to its degree. This degree is of order $(j/i)^\gamma$ for some $0 < \gamma < 1$, see for example [6], and the proportionality factor therefore inverse to order $\sum_{i=1}^{j-1}(j/i)^\gamma \sim j$. Therefore in order to get edge probabilities that match those in the preferential attachment models we take the arrival time of a vertex as its index and take the kernel κ of preferential attachment type

$$\kappa(x,y) = \beta(x \wedge y)^{-\gamma}(x \vee y)^{\gamma-1},$$

for some density parameter $\beta > 0$ and attachment strength parameter $0 < \gamma < 1$. It is easy to check that the graph thus constructed has a power-law degree distribution with exponent

$$\tau = 1 + \frac{1}{\gamma}.$$

Note that because the kernel is homogeneous of index -1, the connection probabilities $p_{i,j}$ do not depend on n. This reflects the dynamic nature of the preferential attachment model. When $\gamma > \frac{1}{2}$ the power-law exponent τ lies between 2 and 3, a range which is often desired when modelling scale-free networks. Precisely in this case the largest component in the graph has macroscopic size no matter how small the edge density, see Dereich et al. [7], or [12]. Our result determines the asymptotic size of this component when the edge density is small.

Theorem 1. *Let $\frac{1}{2} \leq \gamma < 1$ and write $C_1(\mathcal{G}_n)$ for the number of vertices in the largest component. Then we have*

$$\lim_{n \to \infty} \frac{C_1(\mathcal{G}_n)}{n} = \exp\left(-(1-\gamma)\frac{1+o(1)}{\beta}\right),$$

where the limit is taken in probability, the limiting quantity is deterministic and $o(1)$ refers to the asymptotics $\beta \downarrow 0$.

Remarks:

(i) It should be noted that the largest component is surprisingly small. Cohen et al. [4] predict a much larger asymptotic proportion of order $\beta^{1/(\tau-2)}$ for the size of the largest component in scale-free networks with power-law exponent $\tau \in (2,3)$. This prediction is accurate for the inhomogeneous random graph with the same power-law exponent based on the rank-one kernel, $\kappa(x,y) = \beta(xy)^{-\gamma}$, see [3]. The significant difference in the size of the giant after percolation with a small retention parameter between these models may be used in practice to test whether network data credibly originates from a preferential attachment scheme.

(ii) The case $\gamma = \frac{1}{2}$ is treated by Bollobas et al. in [3]. In [13] Riordan describes the observation of the surprisingly small components as *small giant phenomenon*. Note that in the case $\gamma = \frac{1}{2}$ preferential attachment and rank-one kernel coincide.

(iii) Eckhoff et al. [8] have an analogous result for the regime $0 < \gamma < \frac{1}{2}$ showing a decay of order $e^{-c/\sqrt{\beta-\beta_c}}$ with $c = \frac{\pi}{\sqrt{4-8\gamma}}$ when $\beta \downarrow \beta_c > 0$. In that paper a more involved model is considered, but the result can be adapted to our model, see [10] for details.

(iv) The techniques used in [8] and [3] rely on local neighbourhood approximation of the network and analysis of the survival probability of the approximating tree. This method is not suited to our case as the approximating tree has a superexponential growth and is not easy to handle. Instead, we give a constructive proof for the upper bound based on path counting techniques, relying heavily on the fact that these networks are ultra-small, that is typical distances between vertices are only of the order $\log \log n$. For the lower bound we cut-off the small index vertices from our graph, which allows us to use classical theory of inhomogeneous random graphs, as developed in [3].

2 Proof of Theorem 1

As the case $\gamma = \frac{1}{2}$ is covered by Riordan in [13, (4.1)] we assume from now on that $\gamma \in (\frac{1}{2}, 1)$. We start with the upper bound, which we prove by a path counting argument. In Sect. 3 we show that there exists a *core* of highly connected vertices in the network, with bounded diameter independent of n. The core consists of the vertices with index less or equal to $m := m(n) := \sqrt{n} \, (\log n)^{-\alpha}$, for $\alpha = (4-2\gamma)^{-1}$, which satisfy a mild regularity condition, see Definition 1 and the lemmas thereafter. With high probability any giant component contains the core and thus intersects the set $\{1, \ldots, m\}$. Hence we can upper bound the number of vertices in the largest component by m plus the number of (self-avoiding) paths which connect a non-core vertex $i > m$ with the set $\{1, \ldots, m\}$. To count these paths we use two key observations. The first is a bound on the maximum length of a shortest path. Recall the definition of the graph distance of two vertices x, y, namely

$$d_n(x,y) := \min\{N \colon \exists \text{ path } x_0 x_1 \cdots x_{N-1} x_N \text{ in } \mathcal{G}_n, \text{ such that } x_0 = x, x_N = y\}$$

Obviously any path can at most be of length n before it loops into itself, but actual shortest paths are much shorter. Indeed the typical shortest paths in the giant component are no longer than order $\log \log(n)$.

Proposition 1. *Denote by $\mathcal{C}_1(\mathcal{G}_n)$ the largest component in $\mathcal{G}_n$. Then there exists a constant $B > 0$ such that, in probability,*

$$\frac{1}{n^2} \#\Big\{(x,y) \in \mathcal{C}_1(\mathcal{G}_n)^2 \colon d_n(x,y) \geq 4 \frac{\log \log(n)}{\log\left(\frac{\gamma}{1-\gamma}\right)} + B\Big\} \to 0.$$

This result was already shown by Mönch for various preferential attachment models in [11] and the proof for the inhomogeneous random graph of preferential attachment type is deferred to Sect. 3. In the proof we show the following subresult, see Lemmas 6 and 7,

$$\frac{1}{n} \#\Big\{x \in \mathcal{C}_1(\mathcal{G}_n) \colon d_n(x, \mathsf{core}_n) \geq 2 \frac{\log \log(n)}{\log\left(\frac{\gamma}{1-\gamma}\right)} + B\Big\} \to 0.$$

The second observation is that the number of paths connecting to the core is, in expectation, dominated by the paths, which connect the non-core vertices quickly to a powerful vertex, i.e. a vertex with a small index. To count them, define a decreasing cutoff-sequence $(t_l)_{l \in \mathbb{N}_0} \subset (0,1)$ and let $X_l^{(1)}$ be the number of paths $x_0 x_1 \cdots x_{l-1} x_l$ of length l, which satisfy the condition $x_k > t_k n$ for all $0 \leq k < l$ and $x_l \leq t_l n$. These are the paths, which hit an early (i.e. small index) vertex rather fast. The remaining paths stay above the threshold at all times until they arrive at the core. Let $X_l^{(2)}$ be the number of paths of length l, such that $x_k > t_k n$ for all $0 \leq k \leq l$ and $x_l \leq m$. Let $l^* := l^*(n) := \min\{l : t_l n \leq m(n)\}$ and

$$L := L(n) := 2 \frac{\log\log(n)}{\log\left(\frac{\gamma}{1-\gamma}\right)} + 2C + K, \qquad (1)$$

where C, K are the constants from Sect. 3, i.e. $C = \frac{\log(\beta^2)}{\log(\frac{\gamma}{1-\gamma})}$ and K depends neither on n nor on β. Then, by our observations, we can upper bound the number of vertices in the largest component by

$$C_1(\mathcal{G}_n) \leq t_0 n + \sum_{l=1}^{L} X_l^{(1)} + \sum_{l=l^*}^{L} X_l^{(2)} + o(n), \qquad (2)$$

in probability, where the first term includes vertices which are already in the core. Indeed, with high probability almost all vertices in the largest component can be connected to the core by a shortest path of length no longer than L by Lemmas 6 and 7. Note that the second sum starts in l^*, hence whenever we speak about $X_l^{(2)}$ we assume $l \geq l^*$. Now starting with the first sum, we investigate its expectation

$$\mathbb{E}[X_l^{(1)}] \leq \sum_{0 < x_l \leq t_l n} \sum_{t_{l-1}n < x_{l-1} \leq n} \cdots \sum_{t_0 n < x_0 \leq n} \prod_{i=0}^{l-1} \beta(x_i \vee x_{i+1})^{\gamma-1}(x_i \wedge x_{i+1})^{-\gamma}.$$

Note that in the above we would actually get an equality if we restrict the summation variables to be distinct. We define, for all $k \in \{1, \ldots, l\}$,

$$\mu_k(x_k) := \sum_{t_{k-1}n < x_{k-1} \leq n} \cdots \sum_{t_0 n < x_0 \leq n} \prod_{i=0}^{k-1} \underbrace{\beta(x_i \vee x_{i+1})^{\gamma-1}(x_i \wedge x_{i+1})^{-\gamma}}_{=: p(x_i, x_{i+1})},$$

which we can thus express recursively

$$\mu_{k+1}(x) = \sum_{t_k n < y \leq n} \mu_k(y) p(x,y) \quad \forall 1 \leq k < l. \qquad (3)$$

Moreover from the definition of μ_k it immediately follows that

$$\mathbb{E}[X_l^{(1)}] \leq \sum_{0 < x_l \leq t_l n} \mu_l(x_l). \qquad (4)$$

Going forward we use the recursive representation of μ_k to obtain an upper bound. The following statement was already shown in [6], compare with Lemma 1 there.

Lemma 1. *Let $\mu : \{1,\ldots,n\} \to [0,\infty)$ be a function satisfying*

$$\mu(x) \leq \mathbb{1}_{\{x \geq \bar{t}n\}} \psi x^{\gamma-1} + \phi x^{-\gamma}$$

for all $x \in \{1,\ldots,n\}$, for some constants $\psi, \phi > 0$ and some cutoff $\bar{t} \in (0,1)$. Then, for any $t \leq \bar{t}$,

$$\sum_{tn < y \leq n} \mu(y) p(x,y) \leq \mathbb{1}_{\{x > tn\}} \beta \left(\frac{\phi}{2\gamma - 1}(tn)^{1-2\gamma} + \psi \log\left(\frac{1}{t}\right) \right) x^{\gamma-1}$$

$$+ \beta \left(\phi \log\left(\frac{1}{t}\right) + \frac{\psi}{2\gamma - 1} n^{2\gamma-1} \right) x^{-\gamma}.$$

Proof. We have

$$\sum_{tn < y \leq n} \mu(y) p(x,y) = \mathbb{1}_{\{x > tn\}} \sum_{tn < y < x} \mu(y) p(x,y) + \sum_{y = x \vee tn}^{n} \mu(y) p(x,y)$$

$$= \mathbb{1}_{\{x > tn\}} \sum_{tn < y < x} \mu(y) \beta y^{-\gamma} x^{\gamma-1} + \sum_{y = x \vee tn}^{n} \mu(y) \beta y^{\gamma-1} x^{-\gamma}$$

$$\leq \mathbb{1}_{\{x > tn\}} \beta \sum_{tn < y < x} (\psi y^{-1} + \phi y^{-2\gamma}) x^{\gamma-1} + \beta \sum_{y = x \vee tn}^{n} (\psi y^{2(\gamma-1)} + \phi y^{-1}) x^{-\gamma}$$

$$\leq \mathbb{1}_{\{x > tn\}} \beta \left(\psi \log\left(\frac{x}{tn}\right) + \frac{\phi}{2\gamma-1}(tn)^{1-2\gamma} \right) x^{\gamma-1} + \beta \left(\frac{\psi}{2\gamma-1} n^{2\gamma-1} + \phi \log\left(\frac{n}{tn}\right) \right) x^{-\gamma}$$

$$\leq \mathbb{1}_{\{x > tn\}} \beta \left(\psi \log\left(\frac{1}{t}\right) + \frac{\phi}{2\gamma-1}(tn)^{1-2\gamma} \right) x^{\gamma-1} + \beta \left(\frac{\psi}{2\gamma-1} n^{2\gamma-1} + \phi \log\left(\frac{1}{t}\right) \right) x^{-\gamma}.$$

If β is sufficiently small (e.g. $\beta \leq 2\gamma - 1$) we get the following corollary.

$$\sum_{tn < y \leq n} \mu(y) p(x,y) \leq \mathbb{1}_{\{x > tn\}} \left(\phi(tn)^{1-2\gamma} + \psi \beta \log\left(\frac{1}{t}\right) \right) x^{\gamma-1} + \left(\phi \beta \log\left(\frac{1}{t}\right) + \psi n^{2\gamma-1} \right) x^{-\gamma}.$$

Going forward we apply Lemma 1 inductively with $\bar{t} = t_{k-1}$ and $t = t_k$ in every step k. However to this end we first need the starting value

$$\mu_1(x_1) = \sum_{t_0 n < x_0 \leq n} \beta (x_0 \vee x_1)^{\gamma-1} (x_0 \wedge x_1)^{-\gamma}$$

$$\leq \beta \sum_{t_0 n < x_0 \leq n} \left(\mathbb{1}_{\{x_1 > t_0 n\}} x_1^{\gamma-1} x_0^{-\gamma} + x_0^{\gamma-1} x_1^{-\gamma} \right)$$

$$\leq \beta \left(\mathbb{1}_{\{x_1 > t_0 n\}} [\frac{1}{1-\gamma} x_0^{1-\gamma}]_{x_0 = t_0 n}^{n} x_1^{\gamma-1} + [\frac{1}{\gamma} x_0^{\gamma}]_{x_0 = t_0 n}^{n} x_1^{-\gamma} \right)$$

$$\leq \mathbb{1}_{\{x_1 > t_0 n\}} \frac{\beta n^{1-\gamma}}{1-\gamma} x_1^{\gamma-1} + \frac{\beta n^{\gamma}}{\gamma} x_1^{-\gamma}.$$

We factor out n and let $\psi_1 = \frac{\beta n^{-\gamma}}{1-\gamma}$ and $\phi_1 = \frac{\beta n^{\gamma-1}}{\gamma}$ and define for all $k \geq 1$, by the above corollary

$$\psi_{k+1} = \phi_k(t_k n)^{1-2\gamma} + \psi_k \beta \log\left(\frac{1}{t_k}\right)$$

$$\phi_{k+1} = \phi_k \beta \log\left(\frac{1}{t_k}\right) + \psi_k n^{2\gamma-1}$$

It will later come in handy that we define our sequence $(t_k)_{k \in \mathbb{N}}$ such that $t_k n$ is the largest integer satisfying the following relation

$$\frac{1}{1-\gamma}\phi_k(t_k n)^{1-\gamma} \leq \frac{6 t_0^{1-\gamma}}{\pi^2 k^2}. \tag{5}$$

One can see by definition of ϕ_k above, that $\phi_k = O(n^{\gamma-1})$ and thus t_k really only depends on n in the sense that $t_k n$ has to be an integer. Note that by definition of (ϕ_k) this is a recursive definition for $(t_k)_{k \in \mathbb{N}}$, given t_0 which we are still free to choose. Later on we also need an upper bound for the growth of t_k^{-1}. From the definition of t_k it follows immediately that

$$(t_k n + 1)^{1-\gamma} \geq \frac{6(1-\gamma)t_0^{1-\gamma}}{\pi^2 k^2 \phi_k},$$

and hence, using only the definitions of ϕ_k, ψ_k and t_k

$$(t_{k+2} + \frac{1}{n})^{\gamma-1} \leq \frac{\pi^2(k+2)^2}{6(1-\gamma)t_0}\phi_{k+2}n^{1-\gamma}$$

$$\leq \frac{\pi^2(k+2)^2}{6(1-\gamma)t_0}\left(\phi_{k+1}\beta\log\left(\frac{1}{t_{k+1}}\right) + \psi_{k+1}n^{2\gamma-1}\right)n^{1-\gamma}$$

$$\leq \frac{(k+2)^2}{(k+1)^2}t_{k+1}^{\gamma-1}\beta\log\left(\frac{1}{t_{k+1}}\right) + \frac{\pi^2(k+2)^2}{6(1-\gamma)t_0}\left(\phi_k(t_k n)^{1-2\gamma} + \psi_k\beta\log\left(\frac{1}{t_k}\right)\right)n^\gamma$$

$$\leq \frac{(k+2)^2}{(k+1)^2}t_{k+1}^{\gamma-1}\log\left(\frac{1}{t_{k+1}}\right) + \frac{(k+2)^2}{k^2}t_k^{-\gamma} + \frac{\pi^2(k+2)^2}{6(1-\gamma)t_0}\phi_{k+1}\beta\log\left(\frac{1}{t_k}\right)n^{1-\gamma}$$

$$\leq 2\frac{(k+2)^2}{(k+1)^2}t_{k+1}^{\gamma-1}\beta\log\left(\frac{1}{t_{k+1}}\right) + \frac{(k+2)^2}{k^2}t_k^{-\gamma}.$$

Using the above inequality we want to prove the following claim. For all $k \geq 0$ there exists a constant c, independent of β and n, such that

$$t_k^{-1} \leq c \exp\left(\frac{1}{\beta}\sqrt{\frac{\gamma}{1-\gamma}}^k\right). \tag{6}$$

In accordance with the above we pick $t_0 = \varepsilon e^{-1/\beta}$, where ε is just to ensure that $t_0 n$ is an integer. One can easily verify by (5) that t_1, t_2 satisfy inequality (6). For all $k \geq 1$ we have

$$t_{k+2}^{-1} \leq \left(2\frac{(k+2)^2}{(k+1)^2}t_{k+1}^{\gamma-1}\beta\log\left(\frac{1}{t_{k+1}}\right) + \frac{(k+2)^2}{k^2}t_k^{-\gamma}\right)^{\frac{1}{1-\gamma}}$$

$$\leq \left(c_1 \exp\left(\frac{1}{\beta}\sqrt{\frac{\gamma^{k+1}}{(1-\gamma)^{k-1}}}\right)\left(\sqrt{\frac{\gamma}{1-\gamma}}\right)^{k+1} + c_2 \exp\left(\frac{1}{\beta}\sqrt{\frac{\gamma^{k+2}}{(1-\gamma)^k}}\right)\right)^{\frac{1}{1-\gamma}}$$

for some constants c_1 and c_2. Now observe that

$$\exp\left(\frac{1}{\beta}\sqrt{\frac{\gamma^{k+1}}{(1-\gamma)^{k-1}}}\right)\left(\sqrt{\frac{\gamma}{1-\gamma}}\right)^{k+1} \le \tilde{c}\exp\left(\frac{1}{\beta}\sqrt{\frac{\gamma^{k+2}}{(1-\gamma)^k}}\right),$$

for a suitable constant $\tilde{c}$, which proves claim (6). We apply Lemma 1 combined with (4) and (5), and do not forget the n we factored out earlier, to get

$$\mathbb{E}[X_l^{(1)}] \le \sum_{0<x_l\le t_l n}\mu_l(x_l) \le \sum_{0<x_l\le t_l n} n\phi_l x_l^{-\gamma} \le n\frac{1}{1-\gamma}\phi_l(t_l n)^{1-\gamma} \le n\frac{6\,t_0^{1-\gamma}}{\pi^2 l^2}$$

Summing over all l thus gives us

$$\sum_{l\ge 1}\mathbb{E}[X_l^{(1)}] \le n t_0^{1-\gamma} \tag{7}$$

The paths in $X^{(2)}$ stay above the threshold at all times until they reach the core. Thus for any $l \ge l^*$ we can write the expectation as

$$\mathbb{E}[X_l^{(2)}] \le \sum_{0<x_l\le m}\sum_{t_{l-1}n<x_{l-1}\le n}\cdots\sum_{t_0 n<x_0\le n}\prod_{i=0}^{l-1}\beta(x_i\vee x_{i+1})^{\gamma-1}(x_i\wedge x_{i+1})^{-\gamma}$$

$$\le \sum_{0<x_l\le m}\mu_l(x_l).$$

By the definition of L in (1) it follows from our estimation (6) that $l^* = \min\{l : t_l n \le m(n)\}$ is asymptotically equal to L, i.e. $L = l^* + c_1$ for some constant c_1 independent of n. By the bound (6) and (1) we get

$$t_L^{-1} \le c\exp\left(\frac{1}{\beta}\left(\frac{\gamma}{1-\gamma}\right)^{L/2}\right) \le c\,n^{\beta(\sqrt{\frac{\gamma}{1-\gamma}})^K}.$$

Also recall that $m = \sqrt{n}(\log n)^{-\alpha}$, $\alpha > 0$. Hence we find a constant c_2, such that

$$\sum_{l=l^*}^{L}\mathbb{E}[X_l^{(2)}] \le \sum_{l=l^*}^{L}\sum_{0<x_l\le m} n\phi_l x_l^{-\gamma} \le \frac{c_2}{1-\gamma}n^{1-\gamma+\frac{1}{2}(1+\gamma)}\phi_L$$

$$\le \frac{c_3\,t_0^{1-\gamma}}{L^2}n^{\frac{1}{2}(1+\gamma)}t_L^{\gamma-1} \le \frac{c_3\,t_0^{1-\gamma}}{L^2}n^{\beta\tilde{c}+\frac{1}{2}(1+\gamma)} = o(n).$$

The second to last inequality follows by (5) for a suitable constant c_3 and we get that the term is $o(n)$ for β small enough, since $\frac{1}{2}(1+\gamma) < 1$. Thus we use the above together with (7) in (2) to obtain

$$\mathbb{E}[C_1(\mathcal{G}_n)] \le t_0 n + t_0^{1-\gamma}n + o(n)$$

as $n \to \infty$. For $\beta \to 0$ the t_0 term with the lowest exponent dominates, which in this case is $t_0^{1-\gamma}$. Hence by our choice of t_0 in (6) the upper bound in Theorem 1 follows in expectation.

Now we want to conclude from this that the statement also holds with high probability. So from now on let $X_l = X_l^{(1)} + X_l^{(2)}$ be the number of all *core paths* of length l, note that by definition these paths are not necessarily shortest paths and do not necessarily end in the core. For fixed $l \geq 1$ we want to show the concentration of X_l, to this end we will use McDiarmid's inequality in the enhanced version of [5] which we now restate. Consider a function $f \colon \mathcal{Y}_1 \times \cdots \times \mathcal{Y}_N \to \mathbb{R}$, where each $\mathcal{Y}_i$ is a probability space. Moreover, let $G \subseteq \mathcal{Y}_1 \times \cdots \times \mathcal{Y}_N$ be in the domain of f. We call this the *good event* and say that f satisfies a *bounded difference inequality on G* if there exist constants $c_1, \ldots, c_N$ such that, for all $i \in \{1, \ldots, N\}$ and for all $(y_1, \ldots, y_N), (y_1', \ldots, y_N') \in G$ with $y_j = y_j'$ for all $j \neq i$,

$$|f(y_1, \ldots, y_N) - f(y_1', \ldots, y_N')| \leq c_i.$$

For such f, for any independent random variables $Y_1, \ldots, Y_N$ on the spaces $\mathcal{Y}_1, \ldots, \mathcal{Y}_N$ and for all $\delta > \mathbb{P}(G^c) \sum_{i=1}^N c_i$, it holds that

$$\mathbb{P}\big(|f(Y_1, \ldots, Y_N) - \mathbb{E}[f(Y_1, \ldots, Y_N) \mid (Y_1, \ldots, Y_N) \in G]| > \delta\big)$$
$$\leq 2\mathbb{P}(G^c) + 2\exp\left(-\frac{2(\delta - \mathbb{P}(G^c)\sum_{i=1}^N c_i)^2}{\sum_{i=1}^N c_i^2}\right). \tag{8}$$

First we find independent random variables $Y_{\hat{n}_l+1}^{\text{in}}, Y_{\hat{n}_l+1}^{\text{out}}, \ldots, Y_n^{\text{in}}, Y_n^{\text{out}}$ and a function f such that $X_l = f(Y_{t_ln}^{\text{in}}, Y_{t_ln}^{\text{out}}, \ldots, Y_n^{\text{in}}, Y_n^{\text{out}})$, where $\hat{n}_l = t_l n \vee m(n)$, consequently $N = 2(n - \hat{n}_l - 1)$. We choose $Y_i^{\text{in}} \in \{0,1\}^n$ to be the vector corresponding to the incoming connections of vertex i and $Y_i^{\text{out}} \in \{0,1\}^n$ to be the vector corresponding to the outgoing connections of vertex i into the set $\{1, \ldots, \hat{n}_l\}$. Moreover let $\mathcal{Y}_i^{\text{in}}, \mathcal{Y}_i^{\text{out}}$ be their canonical probability spaces. Thus f has to be the number of all core-paths as a function of $Y_{\hat{n}_l+1}^{\text{in}}, Y_{\hat{n}_l+1}^{\text{out}}, \ldots, Y_n^{\text{in}}, Y_n^{\text{out}}$, i.e. let

$$I_n := \{(i_0, \ldots, i_l) \in \{1, \ldots, n\}^{l+1} : i_k > t_k n \, \forall k < l, i_l \leq \hat{n}_l, i_h \neq i_j \, \forall h \neq j\}$$

be the set of all core paths in the full graph and

$$f(Y_{\hat{n}_l+1}^{\text{in}}, Y_{\hat{n}_l+1}^{\text{out}}, \ldots, Y_n^{\text{in}}, Y_n^{\text{out}}) := \sum_{(i_0, \ldots, i_l) \in I_n} Y_{i_{l-1}}^{\text{out}}(i_l) \prod_{j=0}^{l-2} \left(Y_{i_j}^{\text{in}}(i_{j+1}) \vee Y_{i_{j+1}}^{\text{in}}(i_j)\right).$$

We want our good event to be the event that none of the vertices in our graph has untypically many connections, what is untypical for a vertex depends on its index. Therefore we group the vertices into disjoint *layers* $N^{(0)} := \{t_0 n, \ldots, n\}$, $N^{(k)} := \{t_k n, \ldots, t_{k-1} n - 1\}$ for all $k = 1, \ldots, L-1$ and $N^{(L)} := \{1, \ldots, m(n)\}$. Recall L in (1), moreover let $\pi_n \colon \{1, \ldots, n\} \to \{0, \ldots, L\}$ be the function which maps a vertex onto its layer. Then we define our good event to be

$$G := \Big\{ \forall i \in \{\hat{n}_l + 1, \ldots, n\} : |Y_i^{\text{in}}| + |Y_i^{\text{out}}| + \sum_{j=\hat{n}_l+1}^{i-1} Y_j(i) \leq a_{\pi_n(i)}(n) \Big\}$$

for some constants $a_0(n), \ldots, a_L(n)$ depending on n and $|Y_i| := \sum_{j=1}^n Y_i(j)$. We can write $|Y_i^{\text{in}}| = \sum_{j=i+1}^n U_j$ for independent Bernoulli random variables with parameters $p_j = \beta i^{-\gamma} j^{\gamma-1}$ and $|Y_i^{\text{out}}| + \sum_j Y_j(i) = \sum_{j=1}^{t_k n-1} U_j$ with parameter $p_j = \beta j^{-\gamma} i^{\gamma-1}$. We write $q_j = 1 - p_j$ for all j, also note that for all $j \geq t_l n$ we have $p_j \leq \beta i^{-1} =: p$ and $q_j \leq 1 - \beta n^{\gamma-1} i^{-\gamma} =: q$. Thus, by the Markov inequality,

$$\mathbb{P}(|Y_{t_k n}^{\text{in}}| > \tfrac{a_k(n)}{2}) \leq \prod_{j=t_k n+1}^n \mathbb{E}\left[e^{U_j}\right] e^{-a_k(n)} \leq (q + pe)^n e^{-a_k(n)/2}$$

$$\leq \left(1 + \frac{\beta t_k^{-1} e - \beta}{n}\right)^n e^{-a_k(n)/2},$$

which is $o(\frac{1}{n})$ for $a_k(n)/2 := \beta t_k^{-1} e + \log(n)^2$. The same holds for the outgoing connections as

$$\mathbb{P}(|Y_{t_k n}^{\text{out}}| + \sum_{j=\hat{n}_l+1}^{t_k n-1} Y_j(t_k n) > \tfrac{a_k(n)}{2}) \leq \prod_{j=1}^{t_k n} \mathbb{E}\left[e^{U_j}\right] e^{-a_k(n)}$$

$$\leq \prod_{j=1}^{t_k n} \left(1 + \frac{\beta t_k^{\gamma-1} e - \beta}{n^{1-\gamma}}(t_k n)^{1-\gamma} j^{-1}\right) e^{-a_k(n)/2}$$

$$\leq (t_k n + 1) e^{-a_k(n)/2},$$

where we used that $\beta(t_k^{\gamma-1} e - 1)(\frac{t_k n}{n})^{1-\gamma} \leq 1$ for β small enough and that $\prod_{j=1}^n (1 + j^{-1}) = n+1$. Hence we get for the probability that our good event does not occur

$$\mathbb{P}(G^c) = \mathbb{P}(\exists i : |Y_i^{\text{in}}| + |Y_i^{\text{out}}| + \sum_{j=\hat{n}_l+1}^{i-1} Y_j(i) > a_{\pi_n(i)}(n))$$

$$\leq \sum_{k=0}^l (t_{k-1} - t_k) n \left(\mathbb{P}(|Y_{t_k n}^{\text{in}}| > a_k(n)/2) + \mathbb{P}(|Y_{t_k n}^{\text{out}}| + \sum_{j=\hat{n}_l+1}^{t_k n-1} Y_j(t_k n) > a_k(n)/2)\right)$$

$$\leq C n^2 e^{-\log(n)^2}.$$
(9)

Furthermore we have for all $i \in \{1, \ldots, N\}$ and for all $(y_1, \ldots, y_N), (y_1', \ldots, y_N') \in G$ with $y_j = y_j'$ for all $j \neq i$

$$|f(y_1, \ldots, y_N) - f(y_1', \ldots, y_N')| \leq l \prod_{j=0}^l a_j(n) =: c_i.$$

Indeed, since l is fixed, the maximum difference is achieved by setting $y_i' = 0$ and choosing y_i such that $|y_i| = a_{\pi_n(i)}(n)$ or vice versa. Thus the difference above is bounded by the number of core paths in the graph corresponding to $(y_1, \ldots, y_N)$, which involve vertex i. For any path we have at most l places for i and since we work on G any vertex j has at most $a_{\pi_n(j)}(n)$ connections and hence the claim follows. As c_i does not depend on i, we have for the sum

$$2 \sum_{i=t_l n}^{n} \left(l \prod_{j=0}^{l} a_j(n) \right)^2 \leq 2nl^2 \prod_{i=0}^{l} \left(2\beta t_i^{-1} e + 2\log(n)^2 \right)^2$$

$$\leq 2nl^2 \sum_{k=0}^{l} \binom{l}{k} \left(\prod_{i=l-k}^{l} 4\beta^2 e^2 t_i^{-2} \right) (4\log(n))^{4(l-k)}.$$

Now fix $0 < \delta < \frac{1}{2}$. For l large, e.g. close to $L(n)$, the dominating factor in the sum above is the factor

$$\prod_{i=0}^{l} t_i^{-2} = o(t_L^{-r}), \text{ for any } r > 2.$$

Fix such an r then by (1) and (6) we have that $\prod_{i=0}^{l} t_i^{-2} \leq c_1 n^{\beta c_2} = o(n^\delta)$, for suitable constants c_1, c_2 and β small enough. If l is small the sum is dominated by the factor $\log(n)^{4l}$, which is also $o(n^\delta)$. Thus we get that the sum above is $o(n^{1+\delta})$ in any case. This also yields that

$$\mathbb{P}(G^c) \sum_{i=t_l n}^{n} c_i < C n^{3+\delta} e^{-\log(n)^2} \to 0.$$

By the triangle inequality,

$$\mathbb{P}\left(|X_l - \mathbb{E}[X_l]| \geq n^{\frac{1}{2}+\delta} \right)$$
$$\leq \mathbb{P}\left(|X_l - \mathbb{E}[f(Y_{\hat{n}_l+1}^{\text{in}}, \ldots, Y_n^{\text{out}}) \,|\, (Y_{\hat{n}_l+1}^{\text{in}}, \ldots, Y_n^{\text{out}}) \in G]| \geq n^{\frac{1}{2}+\delta}/2 \right)$$
$$+ \mathbb{P}\left(|\mathbb{E}[X_l] - \mathbb{E}[f(Y_{\hat{n}_l+1}^{\text{in}}, \ldots, Y_n^{\text{out}}) \,|\, (Y_{\hat{n}_l+1}^{\text{in}}, \ldots, Y_n^{\text{out}}) \in G]| \geq n^{\frac{1}{2}+\delta}/2 \right).$$
(10)

By McDiarmid's inequality (8) the first term above is bounded from above by

$$2\mathbb{P}(G^c) + 2\exp\left(-\frac{n^{1+2\delta} + o(1)}{2o(n^{1+\delta})} \right) \xrightarrow{n\to\infty} 0.$$

For the second term in (10) note that since we consider the conditional expectation of an event, the event inside the probability is not random. Write $Y = (Y_{\hat{n}_l+1}^{\text{in}}, \ldots, Y_n^{\text{out}})$ for short and note that by (9)

$$\mathbb{E}[X_l] - \mathbb{E}[f(Y)|Y \in G] = \mathbb{E}[f(Y)\mathbf{1}_{Y\in G}] + \mathbb{E}[f(Y)\mathbf{1}_{Y\notin G}] - \mathbb{E}[f(Y)|Y \in G]$$
$$\leq \mathbb{E}[f(Y)|Y \in G] + \mathbb{P}(G^c)\sup_y f(y) - \mathbb{E}[f(Y)|Y \in G]$$
$$\leq C n^{2+l} e^{-\log(n)^2} \xrightarrow{n\to\infty} 0,$$

since $l = O(\log\log n)$ and therefore $X_l = \mathbb{E}[X_l] + o(n/\log\log n)$ with high probability. Summing over l the upper bound in Theorem 1 follows.

For the lower bound we make use of Theorem 10 in [2]. To this end we let $S \subset (0,1]$ be an interval and introduce, for a continuous, symmetric kernel $\kappa: S^2 \to [0,\infty)$, the inhomogeneous random graph $\mathcal{G}_n^\kappa$ obtained by connecting any pair of distinct vertices i, j such that $i/n, j/n \in S$ with probability $\frac{1}{n}\kappa(\frac{i}{n}, \frac{j}{n}) \wedge 1$. Any vertices i with $i/n \notin S$ are not present in the graph $\mathcal{G}_n^\kappa$. Moreover let T_κ be the operator on the space of bounded measurable functions $f: S \to \mathbb{R}$, given by

$$T_\kappa(f)(x) = \int_S \kappa(x,y) f(y) dy.$$

Now we restate said theorem for the readers' convenience.

Lemma 2. *Let $0 \le t < 1$ and $\kappa: [t,1]^2 \to (0,\infty)$ a symmetric continuous function. Let $c > 0$ be a constant and $\phi \in C[0,1]$ strictly positive such that $c T_\kappa(\phi)(x) \ge \phi(x)$, for all $x \in [t,1]$. Let $0 < \varepsilon < 1$ and $c' = (1+\varepsilon)c \le 1$ and define $\mathcal{G}_n^\kappa(c')$ as the graph, which is obtained from $\mathcal{G}_n^\kappa$ by keeping every edge independently with probability c'. Then, with high probability, $\mathcal{G}_n^\kappa(c')$ contains a component of order at least $Cn - o(n)$, where*

$$C = \left(\frac{\varepsilon}{1+\varepsilon}\right) \frac{\int_t^1 \phi(x)\,dx}{\sup_{t \le x \le 1} \phi(x)}.$$

Note that in the original source the authors consider only the interval $[0,1]$, but going through the proof of the theorem reveals that this also holds in the more general form above. Let $\kappa: [t,1]^2 \to (0,\infty)$ be given by $\kappa(x,y) = \beta(x \wedge y)^{-\gamma}(x \vee y)^{\gamma-1}$. Therefore we now need to find a constant $c \in (0,1)$ and a function ϕ such that $c T_\kappa(\phi)(x) \ge \phi(x)$ for all $x \in [t,1]$. We make an educated guess and choose $\phi(x) = x^{\gamma-1}$, then we have

$$T_\kappa(\phi)(x) = \beta x^{\gamma-1} \int_t^x y^{-\gamma} y^{\gamma-1} dy + \beta x^{-\gamma} \int_x^1 y^{\gamma-1} y^{\gamma-1} dy$$

$$= \beta x^{\gamma-1} [\log(y)]_t^x + \beta x^{-\gamma} \frac{1}{2\gamma-1} [y^{2\gamma-1}]_x^1 = \beta x^{\gamma-1} \left[\log(x) - \log(t) + \frac{x^{1-2\gamma}}{2\gamma-1} - \frac{1}{2\gamma-1}\right].$$

By differentiating we get

$$\frac{d}{dx}\left[\log(x) - \log(t) + \frac{x^{1-2\gamma}}{2\gamma-1} - \frac{1}{2\gamma-1}\right] = x^{-1} - x^{-2\gamma} \le 0,$$

since $\gamma > \frac{1}{2}$. Thus the term in brackets is monotone decreasing in x and therefore takes its minimal value at $x = 1$. Hence $T(\phi)(x) \ge \beta \log(\frac{1}{t}) x^{\gamma-1}$. Now we need that $c < 1$, which is equivalent to $\beta \log(\frac{1}{t}) > 1$. That is why we choose our cutoff to be

$$t := \exp\left(-\frac{1+\varepsilon}{\beta}\right),$$

for some $\varepsilon > 0$ arbitrary. Moreover we have

$$\int_t^1 x^{\gamma-1} dx = \frac{1}{\gamma}(1 - t^\gamma) \overset{t \to 0}{\to} \frac{1}{\gamma} \quad \text{and} \quad \sup_{t \le x \le 1} x^{\gamma-1} = t^{\gamma-1}.$$

We apply Lemma 2 with $c = \frac{1}{1+\varepsilon}$, $c' = 1$ and get that we have, with high probability, a giant component of size at least

$$C_1(\mathcal{G}_n) \geq \left(\frac{1}{\gamma} + o(1)\right) \frac{\varepsilon}{1+\varepsilon} \left(t^{\gamma-1}\right)^{-1} n \geq \exp\left(-(1-\gamma)\frac{1+2\varepsilon}{\beta}\right) n,$$

with the last inequality holding for β small enough. Since ε was arbitrary this completes the proof.

3 Proof of Proposition 1

We construct a path from an initial vertex to the core by connecting successively to more powerful vertices (i.e. vertices with lower index). Recall that $m := m(n) := \sqrt{n} (\log n)^{-\alpha}$, for $\alpha = (4-2\gamma)^{-1}$. The core consists of vertices with index $\leq m(n)$, which satisfy a condition on their degree, see below. Inside the core distances are very small, in particular the diameter of the core is bounded by a constant independent of n. For the construction we use the graph consisting of $2n$ vertices, as two powerful vertices in $\mathcal{G}_{2n}$ are typically connected by a vertex in $\mathcal{G}_{2n} \setminus \mathcal{G}_n$. We start by defining the core and proving that it has bounded diameter.

Definition 1.

1. Let $x \neq y \in \{1, \ldots, n\}$ be two vertices in $\mathcal{G}_n$. We call $u \in \{n+1, \ldots, 2n\}$ a *1-connector* for x and y, if $x \leftrightarrow u$ and $u \leftrightarrow y$ in $\mathcal{G}_{2n}$. If x and y are connected by a 1-connector we write $x \xleftrightarrow{1} y$.
2. We define the core as the graph $\text{core}_n := (C_n, E_n)$ with

$$C_n := \{x \in \{1, \ldots, m\} : \deg_{\mathcal{G}_{2n}}(x) - \deg_{\mathcal{G}_n}(x) \geq \sigma\mathbb{E}[\deg_{\mathcal{G}_n}(x)]\},$$

where $\sigma = (1-\gamma)\gamma$. The edges are given by

$$E_n := \left\{(x,y) \in C_n \times C_n : x \xleftrightarrow{1} y\right\}.$$

Note that this is not a subgraph of $\mathcal{G}_{2n}$, however two connected vertices in core_n are also connected in $\mathcal{G}_{2n}$ by a two step connection. Before we prove that core_n has bounded diameter, let us first establish an auxiliary result, which we will use multiple times throughout this section.

Lemma 3.

1. For $x \in \{1, \ldots, n\}$ the probability that x satisfies the degree condition in C_n is uniformly bounded away from 0, i.e. there exists a constant $q > 0$ not depending on n or x such that

$$\mathbb{P}\left(\deg_{\mathcal{G}_{2n}}(x) - \deg_{\mathcal{G}_n}(x) \geq \sigma\mathbb{E}[\deg_{\mathcal{G}_n}(x)]\right) \geq q$$

2. There exists a constant $\rho \in (0,1)$, not depending on n, such that with high probability
$$m \geq \#C_n \geq \rho m.$$

Proof. We first show that there exists a $\theta \in (0,1)$ such that
$$\sigma \mathbb{E}[\deg_{\mathcal{G}_n}(x)] \leq \theta \mathbb{E}[\deg_{\mathcal{G}_{2n}}(x) - \deg_{\mathcal{G}_n}(x)]. \tag{11}$$

We have that
$$\sigma \mathbb{E}[\deg_{\mathcal{G}_n}(x)] \leq \sigma \beta \left(\frac{x^{-\gamma} n^{\gamma}}{\gamma} + \frac{1}{1-\gamma} \right)$$

and also
$$\mathbb{E}[\deg_{\mathcal{G}_{2n}}(x) - \deg_{\mathcal{G}_n}(x)] \geq \int_n^{2n} \beta x^{-\gamma} y^{\gamma-1} \, dy \geq \beta(2^{\gamma} - 1) x^{-\gamma} n^{\gamma}.$$

Hence it sufficient to choose
$$\theta = \sigma(2^{\gamma} - 1)^{-1} \left(\frac{1}{\gamma} + \frac{1}{1-\gamma} \right),$$

which is strictly smaller than 1 by choice of σ. Furthermore let $Z := Z(x) := \deg_{\mathcal{G}_{2n}}(x) - \deg_{\mathcal{G}_n}(x) \geq 0$. It is clear that Z is dominated by a binomial random variable with parameters n and $\tilde{p} = \beta x^{-\gamma} n^{\gamma-1}$ and thus $\mathbb{E}[Z^2] \leq n^2 \tilde{p}^2 + n\tilde{p}$. Therefore by (11), the Paley-Zygmund inequality and our moment estimates for Z, in that order, we have

$$\mathbb{P}\big(\deg_{\mathcal{G}_{2n}}(x) - \deg_{\mathcal{G}_n}(x) \geq \sigma \mathbb{E}[\deg_{\mathcal{G}_n}(x)]\big)$$
$$\geq \mathbb{P}(Z \geq \theta \mathbb{E}[Z]) \geq (1-\theta)^2 \frac{\mathbb{E}[Z]^2}{\mathbb{E}[Z^2]}$$
$$\geq (1-\theta)^2 \frac{(2^{\gamma-1} n\tilde{p})^2}{n^2 \tilde{p}^2 + n\tilde{p}} \geq (1-\theta)^2 \frac{4^{\gamma-1}}{1+\beta^{-1}} =: q,$$

which proves the first statement. For the second, note that the first inequality is trivial. By the first statement $\#C_n$ dominates a binomial random variable with parameters m and q. Hence, by Hoeffding's inequality, for all $t > 0$,
$$\mathbb{P}(\#C_n \leq qm - t) \leq \mathbb{P}(|\text{Bin}(m,q) - qm| \geq t) \leq 2 \exp\left(-\frac{2t^2}{m}\right).$$

For $t = qm/2$ the term above goes to 0 as $m \to \infty$, which implies the second statement for $\rho = q/2$. □

Lemma 4. *For $m = m(n) = \sqrt{n} \log(n)^{-\alpha}$, with $\alpha = \frac{1}{4-2\gamma}$, the graph core_n as defined above has, with high probability, bounded diameter in $\mathcal{G}_{2n}$ independent of n.*

Proof. We prove the statement by comparing core_n with a suitable Erdős-Rényi graph. For $x, y \in C_n$ we want to lower bound the connection probability $\mathbb{P}((x,y) \in E_n)$. Thus let $\mathcal{Z} := \mathcal{Z}(x)$ be the set of all potential 1-connectors for x and y with an edge to x: $\mathcal{Z}(x) := \{u \in \{n+1, \ldots, 2n\} : u \leftrightarrow x\}$. By definition of the core we have that

$$\#\mathcal{Z} \geq \sigma \mathbb{E}[\deg_{\mathcal{G}_n}(x)] \geq \sigma n \beta m^{-\gamma} n^{\gamma-1} = \sigma \beta \left(\frac{n}{m}\right)^\gamma.$$

Hence

$$\mathbb{P}((x,y) \notin E_n) = \prod_{v \in \mathcal{Z}(x)} (1 - \beta y^{-\gamma} v^{\gamma-1}) \leq (1 - \beta m^{-\gamma}(2n)^{\gamma-1})^{\sigma \beta (\frac{n}{m})^\gamma}$$

$$\leq \exp\left(-\sigma \beta^2 \, m^{-2\gamma} 2^{\gamma-1} n^{2\gamma-1}\right) \leq 1 - \frac{\sigma}{2} \beta^2 \, m^{-2\gamma} 2^{\gamma-1} n^{2\gamma-1},$$

where we used that $e^{-z} \leq 1 - z/2$ for z small enough. Therefore

$$\mathbb{P}((x,y) \in E_n) \geq \frac{\sigma}{2} \beta^2 \, m^{-2\gamma} 2^{\gamma-1} n^{2\gamma-1} =: p.$$

Now we use the Erdős-Rényi graph $\mathcal{G}(N,p)$, with $N := \#C_n$ and get $\text{diam}(\mathcal{G}(N,p)) \geq \text{diam}(\mathsf{core}_n)$ with high probability, by construction. To see that $\mathcal{G}(N,p)$ has bounded diameter we use the following result from [1], see Corollary 10 there.

Lemma 5. *Suppose $d = d(N) > 2$ and $0 < p = p(N) < 1$ satisfy $(\log N)/d - 3 \log \log N \to \infty$, $p^d N^{d-1} - 2 \log N \to \infty$ and*

$$(\log N)(p^{d-1} N^{d-2} - \log N + \log \log N) \to -\infty.$$

Then $\mathcal{G}(N,p)$ has, with high probability, diameter $d(N)$.

By Lemma 3 we have that $m \geq \#C_n \geq \rho m$ and thus

$$m = m(n) = \frac{\sqrt{n}}{(\log n)^{\frac{1}{4-2\gamma}}} \quad \text{and} \quad d = \frac{2}{2\gamma - 1}$$

satisfy the assumptions of the above lemma, which completes the proof. $\square$

As mentioned earlier we are aiming to construct a path from any vertex in the giant component into the core. To this end we introduce *layers* of higher and higher connected vertices using a sequence $(t_k)_{k \in \mathbb{N}_0}$ similar to the last section. Here we take inspiration from estimate (6), letting

$$t_k := \delta \exp\left(-\frac{1}{\beta^2}\left(\frac{\gamma}{1-\gamma}\right)^k\right), \quad \text{for some } \delta \in (0,1).$$

Definition 2. 1. For all $0 \leq k < n$ we define the $k - th$ layer in $\mathcal{G}_{2n}$ as

$$\mathcal{N}^{(k)} := \{\lfloor t_{k+1} n \rfloor, \ldots, \lceil t_k n \rceil\}.$$

2. We call a vertex *good* if $x \in \mathcal{N}^{(k)}$ for some $0 \leq k < n$ and
$$\deg_{\mathcal{G}_{2n}}(x) - \deg_{\mathcal{G}_n}(x) \geq \sigma \mathbb{E}[\deg_{\mathcal{G}_n}(x)].$$

We first need to find a good vertex around our initial vertex. In [7] (see also Sect. 2 of [12]) a coupling of the rooted graph $(\mathcal{G}_{2n}, U)$ to a branching random walk $\mathfrak{T}$ is constructed, where U is a distinguished vertex called the *root* chosen uniformly at random from $\mathcal{G}_{2n}$ and $\mathfrak{T}$ is a branching random walk on the negative half-axis with a killing barrier at 0, initial particle at $-X$, where X is standard exponential, and Poisson offspring with intensity measure π given by

$$\pi(\mathrm{d}y) = \beta\big(e^{(1-\gamma)y}\mathbb{1}_{y<0} + e^{\gamma y}\mathbb{1}_{y>0}\big)\mathrm{d}y.$$

The coupling is such that, with high probability, a local neighbourhood exploration of U in $\mathcal{G}_{2n}$ and of $-X$ in $\mathfrak{T}$ up to a finite number of steps yield the same graph. Additionally, particles of $\mathfrak{T}$ at position $x < 0$ are mapped to vertex $i \in \mathcal{G}_{2n}$ iff

$$-\sum_{j=i}^{2n}\frac{1}{j} < x \leq -\sum_{j=i+1}^{2n}\frac{1}{j}.$$

Alternatively one could couple only a finite number of exploration steps and apply the general theorem for inhomogeneous random graphs [9, Theorem 3.14] for the type space $(0, 1]$. One then gets the stated intensity measure above by log transforming the type space to $(-\infty, 0]$.

Lemma 6. *Let $U \in \mathcal{G}_{2n}$ be a uniformly chosen vertex and let $\mathcal{C}_{2n}^K(U)$ denote the local neighbourhood exploration around U after K steps. Then, for every $\varepsilon > 0$ there exists a $K = K(\varepsilon) \in \mathbb{N}$ such that*

$$\mathbb{P}\left(\mathcal{C}_{2n}^K(U) \text{ contains a good vertex}\right) \geq \mathbb{P}(|\mathfrak{T}| = \infty) - \varepsilon,$$

where $\mathbb{P}(|\mathfrak{T}| = \infty)$ is the survival probability of the killed branching random walk.

Proof. We introduce the notation

$$E_k = \{\text{the coupling fails before step } k+1\} \quad \text{for } k \in \mathbb{N}.$$

A good vertex in the graph is coupled to a particle in $\mathfrak{T}$, with a position to the left of $\log(t_0)$ and a certain (large) number of offspring in $[\log(\frac{1}{2}), 0)$. Therefore we can equivalently search for *good particles* in $\mathfrak{T}$. For $k \in \mathbb{N}$ we define $\mathfrak{T}^k$ as the killed branching random walk, which we get after the first k explored vertices in $\mathfrak{T}$. Then

$$\begin{aligned}&\mathbb{P}\left(\mathcal{C}_{2n}^K(U) \text{ contains a good vertex}\right) \\ &\geq \mathbb{P}\left(\mathfrak{T}^K \text{ contains a good particle}\right) - \mathbb{P}(E_K) \\ &\geq \mathbb{P}(|\mathfrak{T}| = \infty)\mathbb{P}\left(\mathfrak{T}^K \text{ contains a good particle} \mid |\mathfrak{T}| = \infty\right) - \mathbb{P}(E_K).\end{aligned} \quad (12)$$

We need the following fact for our argument

$$|\mathfrak{T}| = \infty \iff \forall L < \infty \text{ we have } |\mathfrak{T} \cap (-\infty, -L]| = \infty \text{ almost surely.} \tag{13}$$

The implication $\Leftarrow$ is trivial. The other we prove by contradiction. By survival there are particles in every generation. Assume we have one generation G after which no more offspring in $(-\infty, -L]$ occur, for some $L < \infty$. Let $(y_1, y_2, \ldots)$ denote particles, not necessarily descendants of one another, such that y_i is in generation $G + i$. Then every y_i has positive probability $p = 1 - \exp(-\pi((-\infty, -L)))$ to produce offspring to the left of $-L$. Hence we have

$$\mathbb{P}\left(\forall i \in \mathbb{N} : y_i \text{ has no offspring in } (-\infty, -L)\right) = \prod_{i=1}^{\infty}(1-p) = 0.$$

This is a contradiction, which proves statement (13).

Let $b = \log(t_0)$ and note that for every $M \in \mathbb{N}$ we have for the conditional probability in (12)

$$\mathbb{P}(\mathfrak{T}^K \text{ contains a good particle} \,|\, |\mathfrak{T}| = \infty) \geq \mathbb{P}\left(|\mathfrak{T}^K \cap (-\infty, b)| \geq M \,|\, |\mathfrak{T}| = \infty\right) \cdot$$
$$\mathbb{P}\left(\mathfrak{T}^K \text{ contains a good particle} \,|\, |\mathfrak{T}^K \cap (-\infty, b)| \geq M, |\mathfrak{T}| = \infty\right).$$

By Lemma 3 every particle to the left of $b = \log(t_0)$ has a uniform positive probability to be good. So we find M not depending on K such that

$$\mathbb{P}\left(\mathfrak{T}^K \text{ contains a good particle} \,|\, |\mathfrak{T}^K \cap (-\infty, b)| \geq M, |\mathfrak{T}| = \infty\right) \geq 1 - \tfrac{\delta}{2}.$$

Moreover, by (13) we can pick $K = K(M)$ such that

$$\mathbb{P}\left(|\mathfrak{T}^K \cap (-\infty, b)| \geq M \,|\, |\mathfrak{T}| = \infty\right) \geq 1 - \tfrac{\delta}{2}.$$

Then we get by the above that

$$\mathbb{P}\left(\mathfrak{T}^K \text{ contains a good particle} \,|\, |\mathfrak{T}| = \infty\right) \geq (1 - \tfrac{\delta}{2})^2 \geq 1 - \delta.$$

Now we choose $\delta = \frac{\varepsilon}{2\mathbb{P}(|\mathfrak{T}|=\infty)}$ and thus get from (12) that

$$\mathbb{P}\left(\mathcal{C}_{2n}^K(x) \text{ contains a good vertex}\right) \geq \mathbb{P}(|\mathfrak{T}| = \infty)(1 - \delta) - \mathbb{P}(E_K)$$
$$\geq \mathbb{P}(|\mathfrak{T}| = \infty) - \frac{\varepsilon}{2} - \frac{\varepsilon}{2}.$$

$\square$

As mentioned earlier we now construct a path along good vertices into the core. We let $\mathcal{F}_0$ be the σ-algebra generated by the exploration of a uniformly chosen vertex $U \in \mathcal{G}_n$ stopped either at the *good event* when the first good vertex V is found or at the *bad event* when the coupling fails or we have explored K vertices without finding a good vertex.

Lemma 7. Let $\varepsilon > 0$. There exists a constant $C \leq 0$ such that on the good event

$$\mathbb{P}\left(d_{2n}(V, \mathsf{core}_n) > 2\frac{\log\log(n)}{\log\left(\frac{\gamma}{1-\gamma}\right)} + 2C \,\Big|\, \mathcal{F}_0\right) \leq \varepsilon,$$

if n is sufficiently large.

Proof. We iteratively construct a path from V to the core. Assuming we have found a good vertex $V_k \in \mathcal{N}^{(k)}$ we aim to connect it to a good vertex in $V_{k+1} \in \mathcal{N}^{(k+1)}$ using a 1-connector. The good event ensures that we can start this construction with $V \in \mathcal{N}^{(k)}$, without loss of generality we can assume $k = 0$. Note that in every step this adds two edges to our path. Let $\mathcal{F}_k$ be the σ-algebra generated by our construction up until we found V_k, for $k = 0$ this coincides with $\mathcal{F}_0$ from the conditions of the lemma. Let $\mathcal{Z}(V_k) := \{u \in \{n+1, \ldots, 2n\} : u \leftrightarrow V_k\}$, then we have

$$\mathbb{P}(\nexists \text{ good } w \in \mathcal{N}^{(k+1)} : w \overset{1}{\leftrightarrow} V_k \,|\, \mathcal{F}_k)$$
$$= \mathbb{E}\Big[\prod_{w \in \mathcal{N}^{(k+1)}} \mathbb{P}(\{w \text{ not good}\} \cup \{\nexists i \in \mathcal{Z}(V_k) : i \leftrightarrow w \text{ and } w \text{ good}\}) \,\Big|\, \mathcal{F}_k\Big]$$
$$\leq \mathbb{E}\Big[\prod_{w \in \mathcal{N}^{(k+1)}} \Big(\mathbb{P}(w \text{ not good}) + \mathbb{P}(w \text{ good})\prod_{i \in \mathcal{Z}(V_k)} \mathbb{P}(i \not\leftrightarrow w \,|\, \{w \text{ good}\}, \bigcap_{\substack{j \in \mathcal{Z}(V_k): \\ j < i}} \{j \not\leftrightarrow w\})\Big) \,\Big|\, \mathcal{F}_k\Big].$$

By Lemma 3, $\mathbb{P}(w \text{ good}) \geq q$ and

$$\mathbb{P}\Big(i \not\leftrightarrow w \,\Big|\, \{w \text{ good}\}, \bigcap_{\substack{j \in \mathcal{Z}(V_k): \\ j < i}} \{j \not\leftrightarrow w\}\Big) \leq 1 - \frac{\beta 2^{\gamma-1} t_{k+1}^{-\gamma}}{n}.$$

Plugging this in, we get

$$\mathbb{P}(\nexists \text{ good } w \in \mathcal{N}^{(k+1)} : w \overset{1}{\leftrightarrow} V_k \,|\, \mathcal{F}_k) \leq \Big(1 - q + q\Big(1 - \frac{\beta 2^{\gamma-1} t_{k+1}^{-\gamma}}{n}\Big)^{\sigma \mathbb{E}[\deg_{\mathcal{G}_n}(v)]}\Big)^{|\mathcal{N}^{(k+1)}|}$$
$$\leq \exp\Big(-c\frac{\beta^2 t_{k+1}^{-\gamma}}{n} t_k^{-\gamma}(t_k - t_{k+1})n\Big) \leq \exp\Big(-c\beta^2(t_{k+1}^{-\gamma} t_k^{1-\gamma} - t_{k+1}^{1-\gamma} t_k^{-\gamma})\Big)$$
$$\leq \exp\Big(-c\delta^{1-2\gamma} \tau \left(\frac{\gamma}{1-\gamma}\right)^k\Big),$$

for a suitable constant c, where the last step follows by series expansion of the exponential function. Furthermore the constant in the above is given by

$$\tau = \tau(\gamma) = (1-\gamma)\left(\left(\frac{\gamma}{1-\gamma}\right)^2 - 1\right) > 0.$$

Thus we get by summing over all k

$$\mathbb{P}(\text{the construction fails}) \leq \sum_{k=0}^{\infty} \exp\Big(-c\delta^{1-2\gamma} \tau \left(\frac{\gamma}{1-\gamma}\right)^k\Big) =: \sum_{k=0}^{\infty} a_k.$$

It is clear that the sequence $(\frac{a_{k+1}}{a_k})_{k \in \mathbb{N}_0}$ is strictly monotonically decreasing, with $\frac{a_{k+1}}{a_k} \to 0$ as $k \to \infty$. Hence for $r = \frac{a_1}{a_0}$ we have that $a_{k+1} \leq r a_k$ for all $k \geq 0$. Therefore we get

$$\sum_{k=0}^{\infty} a_k = a_0 + \sum_{k=1}^{\infty} a_k \leq a_0 + \sum_{k=1}^{\infty} r^k a_0 = a_0 + a_0 \frac{r}{1-r}.$$

Since δ was arbitrary, we can choose

$$\delta = \delta(\varepsilon) = \left(\frac{-\log(\frac{\varepsilon}{2})}{c\tau}\right)^{\frac{1}{1-2\gamma}} \in (0,1) \quad \text{for } \varepsilon \text{ small enough.}$$

Thus by the above

$$\mathbb{P}(\text{the construction fails}) \leq \frac{\varepsilon}{2} + \frac{\varepsilon}{2} \frac{r}{1-r} \leq \varepsilon,$$

for $\varepsilon > 0$ arbitrary. Hence we can carry on our construction until we reach the core, that is until $t_L n \leq \sqrt{n} \log(n)^{-\alpha}$ by Lemma 4 and $\alpha = \frac{1}{4-2\gamma}$. This is satisfied by

$$L = L(n) = \frac{\log \log(\sqrt{n} \log(n)^\alpha)}{\log\left(\frac{\gamma}{1-\gamma}\right)} + C, \quad \text{where } C = \frac{\log(\beta^2)}{\log\left(\frac{\gamma}{1-\gamma}\right)}. \tag{14}$$

Recalling the definition of the core, see Definition 1, we see that a good vertex in layer L is also in the core for n large enough. As $\log\log(\sqrt{n}\log(n)^\alpha) = \log\log(n) + \log \frac{1}{2} + o(1)$ this completes the proof. □

Thus we have acquired all the tools necessary to prove Proposition 1. Fix $\varepsilon > 0$. Now let U, V be independent, uniform random variables on $\{1, \ldots, n\}$. Then,

$$\frac{1}{n^2} \#\left\{(x,y) \in \mathcal{C}_1(\mathcal{G}_n)^2 : d_n(x,y) \geq 4 \frac{\log\log(n)}{\log\left(\frac{\gamma}{1-\gamma}\right)} + B\right\}$$

$$= \mathbb{P}\left(U, V \in \mathcal{C}_1(\mathcal{G}_n)^2\right) \mathbb{P}\left(d_n(U,V) \geq 4 \frac{\log\log(n)}{\log\left(\frac{\gamma}{1-\gamma}\right)} + B \,\middle|\, U, V \in \mathcal{C}_1(\mathcal{G}_n)^2\right)$$

The first probability is strictly greater than 0 since $\gamma > 1/2$ and can be bounded from above by 1. Let $\mathcal{F}_0$ be the σ-algebra generated by the exploration around U and V until we are stopped by either the good or bad event. On the good events we found good vertices u and v in at most $K = K(\varepsilon)$ steps. Recalling $L(n)$ from (14) we have

$$\mathbb{P}(d_n(U,V) < 4L(n) + B \mid U, V \in \mathcal{C}_1(\mathcal{G}_n)^2)$$
$$= \mathbb{E}\left[\mathbb{E}\left[\mathbb{P}(d_n(U,V) < 4L(n) + B \mid U, V \in \mathcal{C}_1(\mathcal{G}_n)^2) \middle| \mathcal{F}_0\right]\right]$$
$$\geq \mathbb{E}\left[\mathbb{E}\left[\mathbb{P}\left(\mathcal{C}_{2n}^K(U), \mathcal{C}_{2n}^K(V) \text{ contain good } u,v \mid U, V \in \mathcal{C}_1(\mathcal{G}_n)^2\right) \cdot \right.\right.$$
$$\left.\left. \mathbb{P}(d_n(u,v) < 4L(n) + B \mid \mathcal{C}_{2n}^K(U), \mathcal{C}_{2n}^K(V) \text{ contain good } u,v,\, U, V \in \mathcal{C}_1(\mathcal{G}_n)^2) \middle| \mathcal{F}_0\right]\right].$$

By Lemma 6 the first term is greater or equal $1-\varepsilon$. In the second term since u,v are good they can each be connected to the core in $2L$ steps with probability greater then $1-2\varepsilon$, if n is sufficiently large by Lemma 7. Finally the diameter of the core is bounded by some constant d by Lemma 4 with high probability. Hence $\mathbb{P}(\mathrm{diam}(\mathrm{core}_n) \leq d) \geq 1-\varepsilon$, for n large enough. All in all we have

$$\mathbb{P}\Big(d_n(U,V) \geq 4\frac{\log\log(n)}{\log\left(\frac{\gamma}{1-\gamma}\right)} + 2K + 2C + d \,\Big|\, U,V \in \mathcal{C}_1(\mathcal{G}_n)^2\Big) \leq 4\varepsilon,$$

for all $\varepsilon > 0$, if n is sufficiently large. Therefore

$$\frac{1}{n^2}\#\Big\{(x,y) \in \mathcal{C}_1(\mathcal{G}_n)^2 : d_n(x,y) \geq 4\frac{\log\log(n)}{\log\left(\frac{\gamma}{1-\gamma}\right)} + B\Big\} \to 0.$$

Remark. Together with Theorem 2 in [6] we get that the upper bound in Proposition 1 is sharp.

Acknowledgments. PM is supported by DFG project 444092244 "Condensation in random geometric graphs" within the priority programme SPP 2265.

Disclosure of Interes. The authors have no competing interests to declare that are relevant to the content of this article.

References

1. Bollobás, B.: The diameter of random graphs. Trans. Amer. Math. Soc. **267**, 41–52 (1981)
2. Bollobás, B., Janson, S., Riordan, O.: The phase transition in the uniformly grown random graph has infinite order. Random Struct. Alg. **26**, 1–36 (2005)
3. Bollobás, B., Janson, S., Riordan, O.: The phase transition in inhomogeneous random graphs. Random Struct. Alg. **31**, 116–123 (2007)
4. Cohen, R., ben Avraham, D., Havlin, S.: Percolation critical exponents in scale-free networks. Phys. Rev. E. **66**(036113) (2002)
5. Combes, R.: An extension of McDiarmid's inequality. Preprint arXiv:1511.05240 (2015)
6. Dereich, M., Mönch, C., Mörters, P.: Typical distances in ultrasmall random networks. Adv. Appl. Probab. **44**, 583–601 (2012)
7. Dereich, S., Mörters, P.: Random networks with sublinear preferential attachment: the giant component. Ann. Probab. **41**, 329–384 (2013)
8. Eckhoff, M., Mörters, P., Ortgiese, M.: Near critical preferential attachment networks have small giant components. J. Stat. Phys. **173**, 663–703 (2018)
9. van der Hofstad, R.: Random Graphs and Complex Networks. Volume Two. Cambridge Series in Statistical and Probabilistic Mathematics (2024)
10. Leifhelm, F.: Inhomogeneous random graphs of preferential attachment type: the size of the giant component near criticality. Master thesis, University of Cologne (2024)
11. Mönch, C.: Distances in preferential attachment networks. Ph.D. thesis, University of Bath (2012)

12. Mörters, P.: Tangent graphs. Pure Appl. Funct. Anal. **6**, 1767–1779 (2023)
13. Riordan, O.: The small giant component in scale-free random graphs. Comb. Probab. Comput. **14**, 897–938 (2005)
14. Söderberg, B.: A general formalism for inhomogeneous random graphs. Phys. Rev. E **66**(066121) (2002)

k-Connectivity Threshold for Superpositions of Bernoulli Random Graphs

Daumilas Ardickas, Mindaugas Bloznelis$^{(\boxtimes)}$, and Rimantas Vaicekauskas

Institute of Computer Science, Vilnius University, Vilnius, Lithuania
mindaugas.bloznelis@mif.vu.lt

Abstract. Let $G_1, \ldots, G_m$ be independent identically distributed Bernoulli random subgraphs of the complete graph $\mathcal{K}_n$ having random vertex sets and random edge densities. Assuming that each G_i has a vertex of degree 1 with positive probability, we establish the k-connectivity threshold as $n, m \to +\infty$ for the union $\cup_{i=1}^m G_i$ defined on the vertex set of $\mathcal{K}_n$.

Keywords: k-connectivity threshold · community affiliation graph · complex network

1 Introduction

Let $(X, Q), (X_1, Q_1), \ldots, (X_m, Q_m), \ldots$ be a sequence of independent and identically distributed bivariate random variables taking values in $\{0, 1, 2 \ldots\} \times [0, 1]$. Given n and m, let $G_1 = (\mathcal{V}_1, \mathcal{E}_1), \ldots, G_m = (\mathcal{V}_m, \mathcal{E}_m)$ be independent Bernoulli random subgraphs of the complete graph $\mathcal{K}_n$ having random vertex sets $\mathcal{V}_i \subset \mathcal{V}$ and random edge sets $\mathcal{E}_i$. Here $\mathcal{V} = \{1, \ldots, n\} =: [n]$ denotes the vertex set of $\mathcal{K}_n$. Each $G_i = (\mathcal{V}_i, \mathcal{E}_i)$ is obtained by firstly sampling (X_i, Q_i) and secondly by selecting a subset of vertices $\mathcal{V}_i \subset \mathcal{V}$ of size $|\mathcal{V}_i| = \min\{X_i, n\}$ uniformly at random from the class of subsets of $\mathcal{V}$ of size $\min\{X_i, n\}$ and retaining edges between selected vertices independently at random with probability Q_i. In particular, G_i is a random graph on $\min\{X_i, n\}$ vertices, where every pair of vertices is linked by an edge independently at random with probability Q_i. Note that given i random variables X_i and Q_i do not need to be independent. We study the union graph $G_{[n,m]} = (\mathcal{V}, \mathcal{E})$ with the vertex set $\mathcal{V} = [n]$ and the edge set $\mathcal{E} = \cup_{i \in [m]} \mathcal{E}_i$. One may view $G_{[n,m]}$ as a random network of overlapping communities $G_1, \ldots, G_m$.

Our motivation for studying the random graph $G_{[n,m]}$ is that it represents a null model of the community affiliation graph of [14,15]. We remark that community affiliation graph is a union of independent Bernoulli random graphs (communities), where community memberships are defined by a non-random design. We also mention related random graph models [11,13].

In the parametric regime $m = \Theta(n)$ as $n, m \to +\infty$ the random graph $G_{[n,m]}$ admits an asymptotic degree distribution and non-vanishing global clustering coefficient [4]. Distributions of subgraph counts have been studied in [3,9]. Letting $m/n \to +\infty$ at the rate $m = \Theta(n \ln n)$ one can make $G_{[n,m]}$ connected with a high probability. The connectivity threshold (under various conditions on the distribution of (X, Q)) was studied in [1,2,5,8].

In the present paper we establish the k-connectivity threshold for $G_{[n,m]}$, where $k \geq 2$. We recall that a graph is called k vertex (edge) connected if removal of any $k-1$ vertices (edges) does not make the (remaining) graph disconnected. Before formulating our result, we introduce some notation.

Given integer $x \geq 0$ and number $q \in [0,1]$ we denote by $G(x, q)$ the Bernoulli random graph with the vertex set $[x] = \{1, \ldots, x\}$ and with the edge probability q (any pair of vertices is declared adjacent independently at random with probability q). We denote

$$h(x, q) = 1 - (1 - q)^{(x-1)_+}$$

the probability that vertex 1 is not isolated in $G(x, q)$. We write $(x)_+ = \max\{x, 0\}$ and assign value 1 to the expression 0^0. Note that $h(1, q) = h(0, q) = 0$ for any $q \in [0, 1]$. We denote by $\mathbf{I}_\mathcal{A}$ the indicator function of an event (or set) $\mathcal{A}$. We denote by $\bar{\mathcal{A}}$ the event complement to $\mathcal{A}$. Furthermore, we denote

$$\alpha = \mathbf{E}(Q\mathbf{I}_{\{X \geq 2\}}), \quad \kappa^* = \mathbf{E}(Xh(X, Q)), \quad \tau^* = \mathbf{E}((X)_2 Q(1-Q)^{X-2}),$$
$$\lambda^* = \lambda^*_{n,m,k} = \ln n + (k-1) \ln \tfrac{m}{n} - \tfrac{m}{n} \kappa^*. \tag{1}$$

We are ready to state our result.

Theorem 1. *Let $k \geq 2$ be an integer. Let $n \to +\infty$. Assume that $m = m(n) = \Theta(n \ln n)$. Assume that $\alpha > 0$ and $\tau^* > 0$ and*

$$\mathbf{E}(Xh(X, Q) \ln(1 + X)) < \infty, \tag{2}$$
$$\mathbf{E}(X \min\{1, XQ\} \ln(1 + X)) < \infty, \tag{3}$$
$$\mathbf{E}(X^j Q^{j-1}) < \infty, \quad 2 \leq j \leq k. \tag{4}$$

Then

$$\mathbf{P}\{G_{[n,m]} \text{ is vertex } k\text{-connected}\} \to 1 \quad \text{for} \quad \lambda^*_{n,m,k} \to -\infty, \tag{5}$$
$$\mathbf{P}\{G_{[n,m]} \text{ is edge } k\text{-connected}\} \to 0 \quad \text{for} \quad \lambda^*_{n,m,k} \to +\infty. \tag{6}$$

We remark that since the vertex k-connectivity implies edge k-connectivity, the dichotomy (5), (6) extends to either sort of k-connectivity (edge and vertex connectivity).

We note that $\tau^* > 0$ means that $\mathbf{P}\{G_i$ is not a clique or empty graph$\} > 0$ $\forall i$. In the proof of Theorem 1 we establish (6) by showing that for $\lambda^*_{n,m,k} \to +\infty$ the random graph $G_{[n,m]}$ has a vertex of degree at most $k-1$ with high probability. Hence the k-connectivity threshold for community affiliation graph with randomly assigned community memberships follows a pattern similar to that of the Erdős-Rényi random graph described in the seminal paper [6] (an obstacle to k-connectivity is a vertex of degree at most $k-1$), see also [7,10,12].

2 Proofs

We only prove (6). The proof of (5) will be given in an extended version of the paper. We mention that condition (4) for $j \geq 3$ is used in the proof of (5).

2.1 Proof of (6)

Before the proof we introduce some notation and collect auxiliary facts. $\square$ denotes the end of a proof. We write $G = G_{[n,m]}$ and denote $d(v)$ (respectively $d_i(v)$) the degree of $v \in \mathcal{V}$ in G (respectively G_i). We put $d_i(v) = 0$ for $v \notin \mathcal{V}_i$. Let

$$N_t = \sum_{v \in \mathcal{V}} \mathbf{I}_{\{d(v)=t\}}$$

be the number of vertices of degree t in G. We write $\tilde{X} = \min\{X, n\}$ and $\tilde{X}_i = \min\{X_i, n\}$, for $1 \leq i \leq m$, and denote $\eta_j = \mathbf{E}(X^j Q^{j-1})$, $j = 2, 3, \ldots$,

$$\kappa = \mathbf{E}(\tilde{X}h(\tilde{X}, Q)), \qquad \tau = \mathbf{E}((\tilde{X})_2 Q(1-Q)^{\tilde{X}-2}),$$
$$\mu = \mathbf{E}(Xh(X,Q)\ln(1+X)), \qquad \mu' = \mathbf{E}(X\min\{1, XQ\}\ln(1+X)), \quad (7)$$
$$\lambda_{n,m,k} = \ln n + (k-1)\ln\frac{m}{n} - \frac{m}{n}\kappa. \quad (8)$$

We note that quantities κ and τ depend on n and tend to κ^* and τ^* as $n \to \infty$.

Fact 1. Assume that $\mu < \infty$. Then $0 \leq \kappa^* - \kappa \leq \frac{\mu}{\ln(1+n)}$. Moreover, for $m = \Theta(n \ln n)$ we have $0 \leq \lambda_{n,m,k} - \lambda^*_{n,m,k} = o(1)$ as $n \to +\infty$.

Proof of Fact 1. Denote $\mu_n = \mathbf{E}(Xh(X,Q)\ln(1+X)\mathbf{I}_{\{X>n\}})$. We have

$$\kappa^* - \kappa = \mathbf{E}(Xh(X,Q)\mathbf{I}_{\{X>n\}}) - \mathbf{E}(nh(n,Q)\mathbf{I}_{\{X>n\}})$$
$$\leq \mathbf{E}(Xh(X,Q)\mathbf{I}_{\{X>n\}}) \leq \frac{\mu_n}{\ln(1+n)}.$$

Note that the right side of the first identity is non-negative because $x \to h(x,q)$ is nondecreasing. Hence $0 \leq \kappa^* - \kappa$. Furthermore, the inequality $\mu_n \leq \mu$ implies $\kappa^* - \kappa \leq \frac{\mu}{\ln(1+n)}$. Moreover, $\mu < \infty$ implies $\mu_n = o(1)$. Hence

$$\lambda_{n,m,k} - \lambda^*_{n,m,k} = \frac{m}{n}(\kappa^* - \kappa) \leq \frac{m}{n}\frac{\mu_n}{\ln(1+n)} = o\left(\frac{m}{n}\frac{1}{\ln(1+n)}\right) = o(1).$$

$\square$

Fact 2. We have $|\tau - \tau^*| \leq \mathbf{E}((X)_2 Q \mathbf{I}_{\{X>n\}})$. Consequently, $\eta_2 < \infty$ implies $|\tau - \tau^*| = o(1)$ as $n \to +\infty$.

Proof of Fact 2. We obtain $|\tau - \tau^*| \leq \mathbf{E}\left((X)_2 Q \mathbf{I}_{\{X>n\}}\right)$ from the inequalities

$$\tau^* - \tau = \mathbf{E}\left((X)_2 Q(1-Q)^{X-2}\mathbf{I}_{\{X>n\}}\right) - \mathbf{E}\left((n)_2 Q(1-Q)^{n-2}\mathbf{I}_{\{X>n\}}\right)$$
$$\leq \mathbf{E}\left((X)_2 Q(1-Q)^{X-2}\mathbf{I}_{\{X>n\}}\right) \leq \mathbf{E}\left((X)_2 Q \mathbf{I}_{\{X>n\}}\right),$$
$$\tau - \tau^* = \mathbf{E}\left((n)_2 Q(1-Q)^{n-2}\mathbf{I}_{\{X>n\}}\right) - \mathbf{E}\left((X)_2 Q(1-Q)^{X-2}\mathbf{I}_{\{X>n\}}\right)$$
$$\leq \mathbf{E}\left((n)_2 Q(1-Q)^{n-2}\mathbf{I}_{\{X>n\}}\right) \leq \mathbf{E}\left((X)_2 Q \mathbf{I}_{\{X>n\}}\right).$$

□

The next lemma is the key step in establishing (6).

Lemma 1. *Let $k \geq 2$ be an integer. Let $n, m \to +\infty$. Assume that $m = \Theta(n \ln n)$. Assume that $\alpha > 0$, $\tau^* > 0$, $\eta_2 < \infty$, and $\mu' < \infty$. For $k \geq 3$ we assume, in addition, that $\eta_3 < \infty$. Then for $\lambda_{n,m,k} \to +\infty$ we have $\mathbf{P}\{N_{k-1} \geq 1\} \to 1$. Moreover, for $\lambda_{n,m,k} \to -\infty$ we have $N_t = o_P(1)$ for $t = 0, 1, \ldots, k-1$.*

Proof of (6). In view of Fact 1 we have $\lambda_{n,m,k} \to \pm\infty \Leftrightarrow \lambda^*_{n,m,k} \to \pm\infty$. For $\lambda_{n,m,k} \to +\infty$ Lemma 1 shows that with probability tending to 1 there exists a vertex of degree $k-1$. Hence by removing $k-1$ edges one can make $G_{[n,m]}$ disconnected. □

2.2 Proof of Lemma 1

We begin with an outline of the proof. We say that $u, v \in \mathcal{V}$ are linked by community $G_i = (\mathcal{V}_i, \mathcal{E}_i)$ if $u, v \in \mathcal{V}_i$ and $\mathcal{E}_i$ contains the edge connecting u and v (denoted $\{u, v\} \in \mathcal{E}_i$). The idea of the proof is based on the observation that (in the range of m, n considered) given a vertex of degree $k-1$ it is likely that all of its neighbours are linked to this vertex by different communities. Motivated by this observation, for $v \in \mathcal{V}$, we introduce events

$$\mathcal{D}(v) = \{S(v) = k-1,\ d_i(v) \in \{0,1\}\ \forall i \in [m]\}, \quad \text{where} \quad S(v) = \sum_{i \in [m]} d_i(v).$$

We say that the vertex v has property $\mathcal{D}$ if the event $\mathcal{D}(v)$ occurs. Note that a vertex with property $\mathcal{D}$ has degree (in G) at most $k-1$. Let

$$N'_{k-1} = \sum_{v \in \mathcal{V}} \mathbf{I}_{\mathcal{D}(v)}$$

denote the number of vertices having property $\mathcal{D}$.

We prove Lemma 1 in two steps. We firstly approximate $N_{k-1} = (1 + o_P(1)) N'_{k-1}$ as $n \to +\infty$ and show that $\mathbf{E} N'_{k-1} \to +\infty$ for $\lambda_{n,m,k} \to +\infty$, see Lemma 2 below. Then we establish the concentration $N'_{k-1} = (1 + o_P(1)) \mathbf{E} N'_{k-1}$, see Lemma 3. The proof of Lemma 1 is given at the very end of the section.

In the proof below we use the following simple observations. We have

$$\mathbf{P}\{d_1(1) = 0\} = \mathbf{P}\{1 \notin \mathcal{V}_1\} + \mathbf{P}\{1 \in \mathcal{V}_1, d_1(v_1) = 0\}$$
$$= 1 - \mathbf{E}\frac{\tilde{X}_1}{n} + \mathbf{E}\frac{\tilde{X}_1}{n}(1 - Q_1)^{\tilde{X}_1 - 1} = 1 - \frac{\kappa}{n}.$$

We similarly show that

$$\mathbf{P}\{d_1(1) = 1\} = \frac{\tau}{n}.$$

We observe that these identities imply $\kappa \geq \tau$.

Lemma 2. *Let $k \geq 2$ be an integer. Let $n \to +\infty$. Assume that $m = \Theta(n \ln n)$. Assume that $\tau^* > 0$ and $\eta_2 < \infty$. For $k \geq 3$ we assume, in addition, that $\eta_3 < \infty$. Then*

$$\mathbf{E} N'_{k-1} = (1 + o(1)) \frac{(\tau^*)^{k-1}}{(k-1)!} e^{\lambda_{n,m,k}}, \tag{9}$$

$$\mathbf{E} |N_{k-1} - N'_{k-1}| = o\left(e^{\lambda_{n,m,k}}\right) + o(1). \tag{10}$$

Proof of Lemma 2. Note that random variables $d(v)$, $v \in \mathcal{V}$ are identically distributed and the probabilities $\mathbf{P}\{\mathcal{D}(v)\}$, $v \in \mathcal{V}$, are all equal. Hence

$$\mathbf{E} N'_{k-1} = n \mathbf{P}\{\mathcal{D}(1)\} \quad \text{and} \quad \mathbf{E}|N_{k-1} - N'_{k-1}| \leq n \mathbf{E}|\mathbf{I}_{\{d(1)=k-1\}} - \mathbf{I}_{\mathcal{D}(1)}|. \tag{11}$$

We obtain (9) from the first identity of (11) and the asymptotic formula

$$\mathbf{P}\{\mathcal{D}(1)\} = (1 + o(1)) \left(\frac{m}{n}\right)^{k-1} \frac{(\tau^*)^{k-1}}{(k-1)!} e^{-\frac{m}{n}\kappa} = (1 + o(1)) \frac{1}{n} \frac{(\tau^*)^{k-1}}{(k-1)!} e^{\lambda_{n,m,k}}. \tag{12}$$

Similarly, we derive (10) from the second inequality of (11) and inequalities

$$\mathbf{E}\left|\mathbf{I}_{\{d(1)=k-1\}} - \mathbf{I}_{\{S(1)=k-1\}}\right| = O\left(\frac{m^4}{n^6}\right) + O\left(\frac{m^k}{n^{k+1}} e^{-\frac{m}{n}\kappa}\right), \tag{13}$$

$$\mathbf{E}\left|\mathbf{I}_{\{S(1)=k-1\}} - \mathbf{I}_{\mathcal{D}(1)}\right| = O\left(\left(\frac{m}{n}\right)^{k-2} e^{-\frac{m}{n}\kappa}\right). \tag{14}$$

We note that since $\tau^* > 0$ bound (14) combined with the first relation of (12) implies

$$\mathbf{P}\{S(1) = k - 1\} = (1 + o(1)) \left(\frac{m}{n}\right)^{k-1} \frac{(\tau^*)^{k-1}}{(k-1)!} e^{-\frac{m}{n}\kappa}. \tag{15}$$

In the remaining part of the proof we show (12), (13), (14). Before the proof we introduce some notation. For $K \subset [m]$ we denote $S_K = \sum_{i \in K} d_i(1)$ and introduce events

$$\mathcal{A}_K = \{d_i(1) \geq 1, \forall i \in K\}, \qquad \mathcal{L}_K = \{S_K = k - 1\} \cap \mathcal{A}_K \cap \{S_{[m] \setminus K} = 0\}.$$

Proof of (12). Event $\mathcal{D}(1)$ is the union of mutually disjoint events

$$\mathcal{D}(1) = \bigcup_{K:\,|K|=k-1} \mathcal{L}_K.$$

We have, by symmetry and independence and identical distribution of $d_1(1)$, ..., $d_m(1)$, that

$$\mathbf{P}\{\mathcal{D}(1)\} = \binom{m}{k-1} \mathbf{P}\{S_{\{1,\ldots,k-1\}} = k-1, \mathcal{A}_{\{1,\ldots,k-1\}}\} \mathbf{P}\{S_{\{k,\ldots,m\}} = 0\}$$

$$= \binom{m}{k-1} \left(\mathbf{P}\{d_1(v_1) = 1\}\right)^{k-1} \left(\mathbf{P}\{d_1(v_1) = 0\}\right)^{m-k+1}$$

$$= \binom{m}{k-1} \left(\frac{\tau}{n}\right)^{k-1} \left(1 - \frac{\kappa}{n}\right)^{m-k+1}. \tag{16}$$

Invoking in (16) approximations $\binom{m}{k-1} = \frac{m^{k-1}}{(k-1)!}(1+o(1))$, $\tau = \tau^* + o(1)$, see Fact 2, and the identity, where we use $\ln(1-x) = -x + O(x^2)$ for $x = o(1)$,

$$\left(1 - \frac{\kappa}{n}\right)^{m-k+1} = e^{(m-k+1)\ln\left(1 - \frac{\kappa}{n}\right)} = e^{-\frac{m}{n}\kappa + O\left(\frac{1}{n} + \frac{m}{n^2}\right)} \tag{17}$$

we obtain the first relation of (12). The second one follows from the identity $\frac{m^{k-1}}{n^{k-2}} e^{-\frac{m}{n}\kappa} = e^{-\lambda_{n,m,k}}$.

Proof of (14). We represent $\{S(1) = k-1\}$ by unions of mutually disjoint events

$$\{S(1) = k-1\} = \bigcup_{K \subset [m]:\, 1 \leq |K| = k-1} \mathcal{L}_K = \mathcal{D}(1) \cup \mathcal{R}(1),$$

where

$$\mathcal{R}(1) = \bigcup_{h=1}^{k-2} \bigcup_{K \subset [m],\, |K|=h} \mathcal{L}_K.$$

Hence

$$0 \leq \mathbf{I}_{\{S(1)=k-1\}} - \mathbf{I}_{\mathcal{D}(1)} \leq \mathbf{I}_{\mathcal{R}(1)}. \tag{18}$$

Using symmetry and the independence of $d_1(1), \ldots, d_m(1)$ we evaluate the expectation

$$\mathbf{EI}_{\mathcal{R}(1)} = \mathbf{P}\{\mathcal{R}(1)\} = \sum_{h=1}^{k-2} \binom{m}{h} \mathbf{P}\{S_{[h]} = k-1, \mathcal{A}_{[h]}\} \mathbf{P}\{S_{[m]\setminus[h]} = 0\}$$

$$\leq \sum_{h=1}^{k-2} \binom{m}{h} \left(\frac{\kappa}{n}\right)^h \left(1 - \frac{\kappa}{n}\right)^{m-h}. \tag{19}$$

Here $[h]$ stands for the set $\{1,\ldots,h\}$. In the last inequality we invoked identity

$$\mathbf{P}\{S_{[m]\setminus[h]} = 0\} = \prod_{i\in[m]\setminus[h]} \mathbf{P}\{d_i(1) = 0\} = \left(1 - \frac{\kappa}{n}\right)^{m-h}$$

and used inequalities

$$\mathbf{P}\{S_{[h]} = k-1, \mathcal{A}_{[h]}\} \leq \mathbf{P}\{\mathcal{A}_{[h]}\} = \prod_{i\in[h]} \mathbf{P}\{d_i(1) \geq 1\} = \left(\frac{\kappa}{n}\right)^h.$$

Using $\left(1 - \frac{\kappa}{n}\right)^{m-h} \leq e^{-\frac{\kappa}{n}(m-h)} = (1+o(1))e^{-\frac{m}{n}\kappa}$ and the fact that $n = o(m)$ we upper bound the quantity in (19) by $O\left(\left(\frac{m}{n}\right)^{k-2} e^{-\frac{m}{n}\kappa}\right)$. Now (14) follows from (18), (19).

Proof of (13). Let $\mathcal{B}$ denote the event that vertex 1 is adjacent to some $u \in \mathcal{V}$ in two communities simultaneously,

$$\mathcal{B} = \{\{1,u\} \in \mathcal{E}_i \cap \mathcal{E}_j \quad \text{for some} \quad u \in \mathcal{V} \setminus \{1\} \quad \text{and some} \quad i < j\}.$$

We observe that on the complement event $\bar{\mathcal{B}}$ we have $d(1) = S(1)$. Hence

$$d(1) = (\mathbf{I}_\mathcal{B} + \mathbf{I}_{\bar{\mathcal{B}}})d(1) = \mathbf{I}_\mathcal{B} d(1) + \mathbf{I}_{\bar{\mathcal{B}}} S(1) = S(1) - R_1, \quad R_1 := \mathbf{I}_\mathcal{B}(S(1) - d(1)).$$

Furhermore, since $R_1 = 0$ implies $d(1) = S(1)$ we have

$$\left|\mathbf{I}_{\{d(1)=k-1\}} - \mathbf{I}_{\{S(1)=k-1\}}\right| \leq \mathbf{I}_{\{d(1)=k-1\}}\mathbf{I}_{\{R_1\geq 1\}} + \mathbf{I}_{\{S(1)=k-1\}}\mathbf{I}_{\{R_1\geq 1\}}.$$

Taking the expected values of both sides we obtain

$$\mathbf{E}\left|\mathbf{I}_{\{d(1)=k-1\}} - \mathbf{I}_{\{S(1)=k-1\}}\right|$$
$$\leq \mathbf{P}\{d(1) = k-1, R_1 \geq 1\} + \mathbf{P}\{S(1) = k-1, R_1 \geq 1\} =: p_1 + p_2.$$

To prove (13) we show that

$$p_i = O\left(\frac{m^4}{n^6}\right) + O\left(\frac{m^k}{n^{k+1}} e^{-\frac{m}{n}\kappa}\right), \quad i = 1, 2. \tag{20}$$

We only prove (20) for $i = 1$. For $i = 2$ the proof is much the same.

Let $N(1)$ denote the set of neighbours of vertex 1 in G. For $d(1) = k-1$ we have $|N(1)| = k - 1$. Let $N^*(1) = (u_1^*,\ldots,u_{k-1}^*)$ be a random permutation of elements of $N(1)$. Then $N(1) = \{u_1^*, \cdots, u_{k-1}^*\}$. Let $\gamma_r = \sum_{i\in[m]} \mathbf{I}_{\{\{1,u_r^*\}\in\mathcal{E}_i\}}$ be the number of communities G_i where 1 and u_r^* are adjacent. For $r = 1, 2, \ldots, k-1$ introduce events

$$\mathcal{H}_r = \{\gamma_r \leq 2\}, \quad \mathcal{H}_{r,2} = \{\gamma_r = 2\}, \quad \mathcal{H}_{r*} = \mathcal{H}_{r,2} \cap \{\gamma_j = 1, \forall j \in [k-1] \setminus \{r\}\}.$$

Using the fact that events $\bigcap_{r=1}^{k-1}\mathcal{H}_r$ and $\bigcup_{r=1}^{k-1}\bar{\mathcal{H}}_r$ are complement to each other, we write

$$p_1 = \mathbf{P}\left\{d(1) = k-1, R_1 \geq 1, \bigcap_{r=1}^{k-1}\mathcal{H}_r\right\} + \mathbf{P}\left\{d(1) = k-1, R_1 \geq 1, \bigcup_{r=1}^{k-1}\bar{\mathcal{H}}_r\right\}$$
$$=: I_1 + I_2.$$

Now assume that event $\{R_1 \geq 1\} \cap \left(\bigcap_{r=1}^{k-1} \mathcal{H}_r\right)$ occurs. Then either there is a single γ_r attaining value 2 (while remaining γ_j, with $j \neq r$, attain value 1) or there are (at least) two γ's, say γ_s and γ_t, attaining value 2. Note that the second alternative only makes sense for $k \geq 3$. Consequently,

$$I_1 \leq \mathbf{P}\left\{d(1) = k-1, \bigcup_{r=1}^{k-1} \mathcal{H}_{r*}\right\} + \mathbf{P}\left\{d(1) = k-1, \bigcup_{\{s,t\} \subset [k-1]} \mathcal{H}_{s,2} \cap \mathcal{H}_{t,2}\right\}$$
$$=: I_3 + I_4,$$

where $I_4 = 0$ for $k = 2$. Let us upper bound I_2, I_3, I_4. We have, by the union bound and symmetry,

$$I_2 \leq \sum_{r=1}^{k-1} \mathbf{P}\{d(1) = k-1, \bar{\mathcal{H}}_r\} = (k-1)I_2', \qquad I_2' := \mathbf{P}\{d(1) = k-1, \gamma_1 \geq 3\},$$

$$I_3 = \sum_{r=1}^{k-1} \mathbf{P}\{d(1) = k-1, \mathcal{H}_{r*}\} = (k-1)I_3', \qquad I_3' := \mathbf{P}\{d(1) = k-1, \mathcal{H}_{1*}\},$$

$$I_4 \leq \sum_{\{s,t\} \subset [k-1]} \mathbf{P}\{d(1) = k-1, \mathcal{H}_{s,2} \cap \mathcal{H}_{t,2}\} = \binom{k-1}{2} I_4',$$

$I_4' := \mathbf{P}\{d(1) = k-1, \mathcal{H}_{1,2} \cap \mathcal{H}_{2,2}\}.$

To show (20) we upper bound probabilities I_2', I_3' and I_4'. The bound (20) follows from respective bounds (22), (27) and (28) shown below.

Let us estimate I_2'. The event $\{d(1) = k-1, \gamma_1 \geq 3\}$ implies that for some $u \in V \setminus \{1\}$ and some $\{j_1, j_2, j_3\} \subset [m]$ we have $\{1, u\} \in \mathcal{E}_{j_1} \cap \mathcal{E}_{j_2} \cap \mathcal{E}_{j_3}$. We have, by the union bound,

$$I_2' \leq (n-1)\binom{m}{3}\mathbf{P}\{\{1, u\} \in \mathcal{E}_{j_1} \cap \mathcal{E}_{j_2} \cap \mathcal{E}_{j_3}\} \tag{21}$$

$$= (n-1)\binom{m}{3}\left(\mathbf{E}\left(\frac{(\tilde{X})_2}{(n)_2}Q\right)\right)^3 = O\left(\frac{m^3}{n^5}\right) = o\left(\frac{m^4}{n^6}\right). \tag{22}$$

Let us estimate I_3'. Recall that $d_1(1)$ and $d_2(1)$ denote the degrees of vertex 1 in G_1 and G_2 respectively. Given $v \in V \setminus \{1\}$, integers $s, t \geq 0$, and $\{i, j\} \subset [m]$, introduce events $\mathcal{C} = \{\gamma_r = 1, \, 2 \leq r \leq k-1\}$,

$$\mathcal{B}_{i,j}(v) = \left\{u_1^* = v, \{1, v\} \in \mathcal{E}_i \cap \mathcal{E}_j, \{1, v\} \notin \mathcal{E}_h \; \forall h \in [m] \setminus \{i, j\}\right\} \cap \mathcal{C},$$

$$\mathcal{B}_{s,t}^*(v) = \{d(1) = k-1, \mathcal{B}_{1,2}(v), d_1(1) = 1+s, d_2(1) = 1+t\} \cap \mathcal{C}.$$

Fix $u \in V \setminus \{1\}$. We have, by symmetry,

$$I_3' = \sum_{v \in V \setminus \{1\}} \sum_{\{i,j\} \subset [m]} \mathbf{P}\{d(1) = k-1, \mathcal{B}_{i,j}(v)\}$$

$$= (n-1)\binom{m}{2}\mathbf{P}\{d(1) = k-1, \mathcal{B}_{1,2}(u)\}. \tag{23}$$

Let us evaluate the probability on the right

$$\mathbf{P}\{d(1) = k - 1, \mathcal{B}_{1,2}(u)\} = \sum_{(s,t):\, 0 \le s+t \le k-2} \mathbf{P}\{\mathcal{B}^*_{s,t}(u)\}. \qquad (24)$$

Consider the graph $G^{-\{1,2\}}$ with vertex set $\mathcal{V}$ and edge set $\cup_{j=3}^m \mathcal{E}_j$. Assume that event $\mathcal{B}^*_{s,t}(u)$ occurs. Then the degree of vertex 1 in $G^{-\{1,2\}}$ (denoted $d^{-\{1,2\}}(1)$) equals $k - 2 - s - t$. Moreover, we have $d^{-\{1,2\}}(1) = \sum_{j=3}^m d_j(1)$ (since $\gamma_r = 1$ for $2 \le r \le k-1$). Hence

$$\mathbf{P}\{\mathcal{B}^*_{s,t}(u)\} \le \mathbf{P}\left\{\{1,u\} \in \mathcal{E}_1 \cap \mathcal{E}_2, \sum_{j=3}^m d_j(1) = k - 2 - s - t\right\}$$

$$= \mathbf{P}\{\{1,u\} \in \mathcal{E}_1 \cap \mathcal{E}_2\} \mathbf{P}\left\{\sum_{j=3}^m d_j(1) = k - 2 - s - t\right\}. \qquad (25)$$

Here we used the independence of the random sets $\mathcal{E}_1, \ldots, \mathcal{E}_m$. The first probability of (25)

$$\mathbf{P}\{\{1,u\} \in \mathcal{E}_1 \cap \mathcal{E}_2\} = \left(\mathbf{E}\left(\frac{(\tilde{X})_2}{(n)_2} Q\right)\right)^2 = O(n^{-4}). \qquad (26)$$

For $k - 2 - s - t \ge 1$ the second probability of (25) is evaluated in the same way as the probability $\mathbf{P}\{S(1) = k - 1\}$ in (15). Now we have $S'(1) := \sum_{j=3}^m d_j(1)$ instead of $S(1) = \sum_{j=1}^m d_j(1)$ and we have $h = k - 2 - s - t$ instead of $k - 1$. For $h = 1, \ldots, k - 2$ the argument of the proof of (15) applies to $\mathbf{P}\{S'(1) = h\}$ and we have

$$\mathbf{P}\{S'(1) = h\} = (1 + o(1)) \left(\frac{m}{n}\right)^h \frac{(\tau^*)^h}{h!} e^{-\frac{m}{n}\kappa}.$$

Furthermore, for $h = 0$ we have

$$\mathbf{P}\{S'(1) = 0\} = (\mathbf{P}\{d_3(1) = 0\})^{m-2} = \left(1 - \frac{\kappa}{n}\right)^{m-2} = (1 + o(1)) e^{-\frac{m}{n}\kappa}.$$

Next, using the fact that $\max_{0 \le h \le k-2} \frac{m^h}{n^h} = \frac{m^{k-2}}{n^{k-2}}$ we obtain the bound

$$\mathbf{P}\{S'(1) = h\} = O\left(\left(\frac{m}{n}\right)^{k-2} e^{-\frac{m}{n}\kappa}\right), \qquad h = 0, 1, \ldots, k-2.$$

Hence the second probability of (25) is $O\left(\left(\frac{m}{n}\right)^{k-2} e^{-\frac{m}{n}\kappa}\right)$. Combining this bound with (26) we obtain $\mathbf{P}\{\mathcal{B}^*_{s,t}(u)\} = O\left(\frac{m^{k-2}}{n^{k+2}} e^{-\frac{m}{n}\kappa}\right)$. Now (24) implies the bound

$$\mathbf{P}\{d(1) = k - 1, \mathcal{B}_{1,2}(u)\} = O\left(\frac{m^{k-2}}{n^{k+2}} e^{-\frac{m}{n}\kappa}\right).$$

Finally, (23) implies the bound

$$I'_3 = O\left(\frac{m^k}{n^{k+1}} e^{-\frac{m}{n}\kappa}\right). \qquad (27)$$

Let us estimate I'_4. Assume that event $\{d(1) = k-1\} \cap \mathcal{H}_{1,2} \cap \mathcal{H}_{2,2}$ occurs. Then for some $\{u,v\} \subset \mathcal{V} \setminus \{1\}$ one of the following alternatives holds:

$\mathcal{A}_1$: for some $i_1 \neq i_2$ we have $\{1,u\}, \{1,v\} \in \mathcal{E}_{i_1} \cap \mathcal{E}_{i_2}$;

$\mathcal{A}_2$: for some $i_1 \neq i_2 \neq i_3$ we have $\{1,u\}, \{1,v\} \in \mathcal{E}_{i_1}$ and $\{1,u\} \in \mathcal{E}_{i_2}$, and $\{1,v\} \in \mathcal{E}_{i_3}$;

$\mathcal{A}_3$: for some $i_1 \neq i_2 \neq i_3 \neq i_4$ we have $\{1,u\} \in \mathcal{E}_{i_1} \cap \mathcal{E}_{i_2}$ and $\{1,v\} \in \mathcal{E}_{i_3} \cap \mathcal{E}_{i_4}$.

Given $\{u,v\}$, we estimate the probabilities $\mathbf{P}\{\mathcal{A}_i\}$, $1 \leq i \leq 3$, using the union bound and symmetry,

$$\mathbf{P}\{\mathcal{A}_1\} \leq \binom{m}{2} \mathbf{P}\{\{1,u\},\{1,v\} \in \mathcal{E}_{i_1}\} \mathbf{P}\{\{1,u\},\{1,v\} \in \mathcal{E}_{i_2}\}$$

$$= \binom{m}{2} \left(\mathbf{E}\left(\frac{(\tilde{X})_3}{(n)_3} Q^2\right)\right)^2,$$

$$\mathbf{P}\{\mathcal{A}_2\} \leq (m)_3 \left(\mathbf{E}\left(\frac{(\tilde{X})_3}{(n)_3} Q^2\right)\right) \left(\mathbf{E}\left(\frac{(\tilde{X})_2}{(n)_2} Q\right)\right)^2,$$

$$\mathbf{P}\{\mathcal{A}_3\} \leq \binom{m}{2}\binom{m-2}{2}\left(\mathbf{E}\left(\frac{(\tilde{X})_2}{(n)_2} Q\right)\right)^4.$$

Furthermore, taking into account that there are $(n-1)_2$ ways to choose vertices $u \neq v$, we have

$$I'_4 \leq (n-1)_2 \left(\mathbf{P}\{\mathcal{A}_1\} + \mathbf{P}\{\mathcal{A}_2\} + \mathbf{P}\{\mathcal{A}_3\}\right)$$

$$= O\left(\frac{m^2}{n^4} + \frac{m^3}{n^5} + \frac{m^4}{n^6}\right) = O\left(\frac{m^4}{n^6}\right). \qquad (28)$$

The latter bound combined with (21) and (27) yields (20). $\square$

Lemma 3. *Let $k \geq 2$ be an integer. Let $n \to +\infty$. Assume that $m = \Theta(n \ln n)$. Assume that $\alpha > 0$, $\tau^* > 0$, $\eta_2 < \infty$, $\mu' < \infty$. Assume that $\lambda_{n,m,k} \to +\infty$. Then $N'_{k-1} = (1 + o_P(1)) \mathbf{E} N'_{k-1}$.*

In the proof of Lemma 3 we use the following fact.

Fact 3. *Let $n \to +\infty$. For $m = \Theta(n \ln n)$ condition $\mu' < \infty$ implies*

$$\mathbf{E}\left((\tilde{X})_2 \left(1 - (1-Q)^{\tilde{X}-1}\right)\right) = o(n^2/m). \qquad (29)$$

Proof of Fact 3. Inequalities $1 - (1-q)^x \leq 1$ and $1 - (1-q)^x \leq qx$ imply inequality $(1 - (1-q)^x \leq \min\{1, qx\}$. From the latter inequality we obtain

$$\mathbf{E}\left((\tilde{X})_2 \left(1 - (1-Q)^{\tilde{X}-1}\right)\right) \leq \mathbf{E}(\tilde{X}^2 \min\{1, \tilde{X}Q\}) =: I.$$

We will show that $I = o(n/\ln n)$. We split

$$I = \mathbf{E}\left(\tilde{X}^2 \min\{1, \tilde{X}Q\}\mathbf{I}_{\{X<\sqrt{n}\}}\right) + \mathbf{E}\left(\tilde{X}^2 \min\{1, \tilde{X}Q\}\mathbf{I}_{\{X\geq\sqrt{n}\}}\right) =: I_1 + I_2.$$

Using $x/\ln(1+x) \leq \sqrt{n}/\ln(1+\sqrt{n})$ for $x < \sqrt{n}$ and $x/\ln(1+x) \leq n/\ln(1+n)$ for $x \leq n$ we upper bound

$$I_1 \leq \frac{\sqrt{n}}{\ln(1+\sqrt{n})}\mathbf{E}\left(X\min\{1, XQ\}\ln(1+X)\mathbf{I}_{\{X<\sqrt{n}\}}\right) \leq \frac{\sqrt{n}}{\ln(1+\sqrt{n})}\mu',$$

$$I_2 \leq \frac{n}{\ln(1+n)}\mathbf{E}\left(\tilde{X}\min\{1, \tilde{X}Q\}\ln(1+\tilde{X})\mathbf{I}_{\{X\geq\sqrt{n}\}}\right) \leq \frac{n}{\ln(1+n)}I_2',$$

where

$$I_2' = \mathbf{E}\left(X\min\{1, XQ\}\ln(1+X)\mathbf{I}_{\{X\geq\sqrt{n}\}}\right).$$

Our condition $\mu' < \infty$ implies $I_2' = o(1)$ as $n \to +\infty$. Hence

$$I \leq \frac{\sqrt{n}}{\ln(1+\sqrt{n})}\mu' + \frac{n}{\ln(1+n)}I_2' = o(n/\ln n).$$

$\square$

Proof of Lemma 3. To show the concentration of N_{k-1}' around the mean value $\mathbf{E}N_{k-1}'$ we upper bound the variance of N_{k-1}'. To this aim we evaluate the covariances $\mathbf{Cov}(I_{\mathcal{D}(v)}, I_{\mathcal{D}(u)})$.

Given vertex $v \in \mathcal{V}$ and set $K \subset [m]$ of size $|K| = k-1$ denote the event

$$\mathcal{D}_K(v) = \{d_i(v) = 1 \ \forall i \in K \text{ and } d_j(v) = 0 \ \forall j \in [m] \setminus K\}.$$

Note that for $K \neq K'$ events $\mathcal{D}_K(v), \mathcal{D}_{K'}(v)$ are mutually disjoint. Hence

$$\mathbf{I}_{\mathcal{D}(v)} = \sum_{K \subset [m], |K|=k-1} \mathbf{I}_{\mathcal{D}_K(v)}.$$

For $h = 0, 1, \ldots, k-1$ we denote $K(h) = \{h+1, \ldots, h+k-1\}$. Observe that $K(0)$ and $K(h)$ share $|K(0) \cap K(h)| = k-1-h$ common elements. We have, by symmetry,

$$\mathbf{E}(\mathbf{I}_{\mathcal{D}(v)}\mathbf{I}_{\mathcal{D}(u)}) = \binom{m}{k-1}\mathbf{E}\left(\mathbf{I}_{\mathcal{D}_{K(0)}(v)} \sum_{K \subset [m], |K|=k-1} \mathbf{I}_{\mathcal{D}_K(u)}\right)$$

$$= \binom{m}{k-1}\sum_{h=0}^{k-1}\binom{k-1}{k-1-h}\binom{m-k+1}{h}\mathbf{E}(\mathbf{I}_{\mathcal{D}_{K(0)}(v)}\mathbf{I}_{\mathcal{D}_{K(h)}(u)}). \quad (30)$$

Let us evaluate $\mathbf{E}(\mathbf{I}_{\mathcal{D}_{K(0)}(v)}\mathbf{I}_{\mathcal{D}_{K(h)}(u)}) = \mathbf{P}\{\mathcal{D}_{K(0)}(v) \cap \mathcal{D}_{K(h)}(u)\}$. To this aim we write event $\mathcal{D}_{K(0)}(v) \cap \mathcal{D}_{K(h)}(u)$ in the form

$$\mathcal{X}_{K(0) \cap K(h)} \cap \mathcal{Y}_{[m] \setminus (K(0) \cup K(h))} \cap \mathcal{Z}_{K(0) \setminus K(h)} \cap \mathcal{W}_{K(h) \setminus K(0)},$$

where for any $A \subset [m]$ we denote events

$$\mathcal{X}_A = \{d_i(v) = d_i(u) = 1 \ \forall i \in A\}, \qquad \mathcal{Y}_A = \{d_i(v) = d_i(u) = 0 \ \forall i \in A\},$$
$$\mathcal{Z}_A = \{d_i(v) = 1, d_i(u) = 0 \ \forall i \in A\}, \qquad \mathcal{W}_A = \{d_i(v) = 0, d_i(u) = 1 \ \forall i \in A\}.$$

By the independence and identical distribution of $G_1, \ldots, G_m$, we have

$$\begin{aligned}\mathbf{P}\{\mathcal{D}_{K(0)}(v) \cap \mathcal{D}_{K(h)}(u)\} &= \mathbf{P}\{\mathcal{X}_{K(0) \cap K(h)}\} \times \mathbf{P}\{\mathcal{Y}_{[m] \setminus (K(0) \cup K(h))}\} \\ &\quad \times \mathbf{P}\{\mathcal{Z}_{K(0) \setminus K(h)}\} \times \mathbf{P}\{\mathcal{W}_{K(h) \setminus K(0)}\} \\ &= q_1^{k-1-h} q_2^{m-k-h+1} q_3^{2h}.\end{aligned} \qquad (31)$$

Here we denote

$$q_1 = \mathbf{P}\{d_1(v) = d_1(u) = 1\}, \qquad q_2 = \mathbf{P}\{d_1(v) = d_1(u) = 0\},$$
$$q_3 = \mathbf{P}\{d_1(v) = 1, d_1(u) = 0\}.$$

We show below that

$$q_1 \leq \frac{1}{n} \mathbf{E}\left(\tilde{X}_1^2 Q_1\right), \qquad q_3 \leq \frac{\tau}{n}, \qquad q_2 = 1 - 2\frac{\kappa}{n} + \frac{\Delta_2}{(n)_2}, \qquad (32)$$

where Δ_2 satisfies $0 \leq \Delta_2 \leq \mathbf{E}\left((\tilde{X}_1)_2 \left(1 - (1 - Q_1)^{\tilde{X}_1 - 1}\right)\right)$.

Using (32) we upper bound the product in (31). Firstly, combining $1 + a \leq e^a$, (32) and (29) we estimate

$$q_2^m \leq \left(e^{-2\frac{\kappa}{n} + \frac{\Delta_2}{(n)_2}}\right)^m = e^{-2\kappa \frac{m}{n}} \left(1 + O\left(\Delta_2 \frac{m}{(n)_2}\right)\right) = (1 + o(1)) e^{-2\kappa \frac{m}{n}}.$$

This bound extends to q_2^{m-t} for small t. In particular, for $0 \leq t \leq 2k-2$, we have

$$\begin{aligned}q_2^{m-t} = (q_2^m)^{1 - \frac{t}{m}} &\leq (1 + o(1)) \, e^{-2\kappa \frac{m}{n}(1 - \frac{t}{m})} \\ &= (1 + o(1)) \left(1 + O(n^{-1})\right) e^{-2\kappa \frac{m}{n}} \\ &= (1 + o(1)) e^{-2\kappa \frac{m}{n}}.\end{aligned} \qquad (33)$$

Now from (32), (33) and (31) we obtain

$$\mathbf{P}\{\mathcal{D}_{K(0)}(v) \cap \mathcal{D}_{K(k-1)}(u)\} = q_2^{m-2k+2} q_3^{2k-2} \leq (1 + o(1))) e^{-2\kappa \frac{m}{n}} \left(\frac{\tau}{n}\right)^{2k-2},$$
$$\mathbf{P}\{\mathcal{D}_{K(0)}(v) \cap \mathcal{D}_{K(h)}(u)\} = O(n^{1-k-h}) e^{-2\kappa \frac{m}{n}}, \qquad h = 0, 1, \ldots, k - 2.$$

Invoking these bounds in (30) and using $\frac{m^h}{n^h} = o\left(\frac{m^{k-1}}{n^{k-1}}\right)$ for $0 < h \leq k - 2$ we have that

$$\begin{aligned}\mathbf{E}(\mathbf{I}_{\mathcal{D}(v)} \mathbf{I}_{\mathcal{D}(u)}) &\leq (1 + o(1)) \binom{m}{k-1} \binom{m-k+1}{k-1} e^{-2\kappa \frac{m}{n}} \left(\frac{\tau}{n}\right)^{2k-2} \\ &= (1 + o(1)) \left(\mathbf{P}\{\mathcal{D}(v)\}\right)^2.\end{aligned} \qquad (34)$$

In the last step we used $\tau = (1 + o(1))\tau^*$ and (12). It follows from (34) that

$$\begin{aligned}\mathbf{Var}N'_{k-1} &= \mathbf{E}(N'_{k-1})^2 - (\mathbf{E}N'_{k-1})^2 \\ &= n\mathbf{P}\{\mathcal{D}(v)\} + (n)_2\mathbf{E}\big(\mathbf{I}_{\mathcal{D}(v)}\mathbf{I}_{\mathcal{D}(u)}\big) - (n\mathbf{P}\{\mathcal{D}(v)\})^2 \\ &\leq n\mathbf{P}\{\mathcal{D}(v)\} + o(1)\left(n\mathbf{P}\{\mathcal{D}(v)\}\right)^2 \\ &= \mathbf{E}N'_{k-1} + o(\mathbf{E}N'_{k-1})^2.\end{aligned}$$

In the case where $\mathbf{E}N'_{k-1} \to +\infty$ we obtain $\mathbf{Var}N'_{k-1} = o(\mathbf{E}N'_{k-1})^2$ as $n \to +\infty$. Now Chebyshev's inequality shows for any $\varepsilon > 0$

$$\mathbf{P}\{|N'_{k-1} - \mathbf{E}N'_{k-1}| > \varepsilon\mathbf{E}N'_{k-1}\} \leq \frac{\mathbf{Var}N'_{k-1}}{(\varepsilon\mathbf{E}N'_{k-1})^2} = \frac{o(1)}{\varepsilon^2} = o(1).$$

Hence $N'_{k-1} = (1 + o_P(1))\mathbf{E}N'_{k-1}$.

It remains to show (32). Let $\mathbf{P}_X$ denote the conditional probability given X_1. Let us estimate q_1. Identities $d_1(v) = 1$, $d_1(u) = 1$ imply $\{u,v\} \subset \mathcal{V}_1$. In particular, we have $q_1 \leq \mathbf{P}\{\{u,v\} \subset \mathcal{V}_1\} = (n)_2^{-1}\mathbf{E}(\tilde{X}_1)_2$. Furthermore, we have

$$\begin{aligned}q_1 &= \mathbf{E}\mathbf{P}_X\{d_1(v) = d_1(v) = 1, \{u,v\} \subset \mathcal{V}_1\} \\ &= \mathbf{E}\left(\mathbf{P}_X\{d_1(v) = d_1(v) = 1|\{u,v\} \subset \mathcal{V}_1\}\mathbf{P}_X\{\{u,v\} \subset \mathcal{V}_1\}\right) \\ &= \mathbf{E}\left(\left(Q_1(1-Q_1)^{2(\tilde{X}_1-2)} + (1-Q_1)((\tilde{X}_1-2)Q_1(1-Q_1)^{\tilde{X}_1-3})^2\right)\frac{(\tilde{X}_1)_2}{(n)_2}\right).\end{aligned}$$

Here the first term $Q_1(1-Q_1)^{2(\tilde{X}_1-2)}$ refers to the event $\{u,v\} \in \mathcal{E}_1$. The second term refers to the complement event $\{u,v\} \notin \mathcal{E}_1$.

Next for $q \in [0,1]$ and $x = 2, 3, \ldots$ we apply inequalities $(x-2)q(1-q)^{x-3} \leq 1$ and $1 - q \leq 1$ and derive the inequality

$$q(1-q)^{2(x-2)} + (1-q)\left((x-2)q(1-q)^{x-3}\right)^2 \leq q + (x-2)q = (x-1)q.$$

Invoking this inequality in the formula for q_1 above we obtain

$$q_1 \leq (n)_2^{-1}\mathbf{E}\left((\tilde{X}_1 - 1)(\tilde{X}_1)_2 Q_1\right) \leq n^{-1}\mathbf{E}\left(\tilde{X}_1^2 Q_1\right).$$

Let us evaluate q_2. We split

$$q_2 = q_{2,1} + q_{2,2} + q_{2,3} + q_{2,4}, \tag{35}$$

where

$q_{2,1} = \mathbf{P}\{\{u,v\} \cap \mathcal{V}_1 = \emptyset\}$, $\quad q_{2,2} = \mathbf{P}\{\{u,v\} \subset \mathcal{V}_1, d_1(v) = d_1(u) = 0\}$,
$q_{2,3} = \mathbf{P}\{v \in \mathcal{V}_1, u \notin \mathcal{V}_1, d_1(v) = 0\}$, $\quad q_{2,4} = \mathbf{P}\{v \notin \mathcal{V}_1, u \in \mathcal{V}_1, d_1(u) = 0\}$,

and calculate the probabilities

$$q_{2,1} = 1 - \mathbf{P}\{\{u \in \mathcal{V}_1\} \cup \{v \in \mathcal{V}_1\}\}$$
$$= 1 - \mathbf{P}\{u \in \mathcal{V}_1\} - \mathbf{P}\{v \in \mathcal{V}_1\} + \mathbf{P}\{\{u,v\} \subset \mathcal{V}_1\}$$
$$= 1 - 2\frac{\mathbf{E}\tilde{X}_1}{n} + \frac{\mathbf{E}(\tilde{X}_1)_2}{(n)_2},$$
$$q_{2,2} = \mathbf{E}\mathbf{P}_X\{\{u,v\} \subset \mathcal{V}_1, d_1(v) = d_1(u) = 0\}$$
$$= \mathbf{E}(\mathbf{P}_X\{d_1(v) = d_1(u) = 0|\{u,v\} \subset \mathcal{V}_1\}\mathbf{P}_X\{\{u,v\} \subset \mathcal{V}_1\}) \quad (36)$$
$$= \mathbf{E}\left((1-Q_1)^{2\tilde{X}_1-3}\frac{(\tilde{X}_1)_2}{(n)_2}\right),$$
$$q_{2,3} = q_{2,4} = \mathbf{E}\mathbf{P}_X\{v \notin \mathcal{V}_1, u \in \mathcal{V}_1, d_1(u) = 0\}$$
$$= \mathbf{E}(\mathbf{P}_X\{d_1(u) = 0|v \notin \mathcal{V}_1, u \in \mathcal{V}_1\}\mathbf{P}_X\{v \notin \mathcal{V}_1, u \in \mathcal{V}_1\})$$
$$= \mathbf{E}\left((1-Q_1)^{\tilde{X}_1-1}\left(1 - \frac{\tilde{X}_1}{n}\right)\frac{\tilde{X}_1}{n-1}\right)$$
$$= \frac{1}{n}\mathbf{E}\left(\tilde{X}_1(1-Q_1)^{\tilde{X}_1-1}\right) - \frac{1}{(n)_2}\mathbf{E}\left((\tilde{X}_1)_2(1-Q_1)^{\tilde{X}_1-1}\right). \quad (37)$$

Invoking these expressions for $q_{2,1}, q_{2,2}, q_{2,3}, q_{2,4}$ in (35) we obtain $q_2 = 1 - 2\frac{\kappa}{n} + \frac{\Delta_2}{(n)_2}$, where

$$\Delta_2 := \mathbf{E}\left((\tilde{X}_1)_2\left(1 - 2(1-Q_1)^{\tilde{X}_1-1} + (1-Q_1)^{2\tilde{X}_1-3}\right)\right)$$
$$\leq \mathbf{E}\left((\tilde{X}_1)_2\left(1 - (1-Q_1)^{\tilde{X}_1-1}\right)\right).$$

Let us evaluate q_3. We split $q_3 = q_{3,1} + q_{3,2}$, where

$$q_{3,1} = \mathbf{P}\{d_1(v) = 1, u \notin \mathcal{V}_1\}, \qquad q_{3,2} = \mathbf{P}\{d_1(v) = 1, u \in \mathcal{V}_1, d_1(u) = 0\},$$

and calculate the probabilities

$$q_{3,1} = \mathbf{E}\mathbf{P}_X\{d_1(v) = 1, u \notin \mathcal{V}_1, v \in \mathcal{V}_1\}$$
$$= \mathbf{E}(\mathbf{P}_X\{d_1(v) = 1|u \notin \mathcal{V}_1, v \in \mathcal{V}_1\}\mathbf{P}_X\{u \notin \mathcal{V}_1, v \in \mathcal{V}_1\})$$
$$= \mathbf{E}\left(Q_1(1-Q_1)^{\tilde{X}_1-2}(\tilde{X}_1-1)\left(1-\frac{\tilde{X}_1}{n}\right)\frac{\tilde{X}_1}{n-1}\right)$$
$$= \mathbf{E}\left(Q_1(1-Q_1)^{\tilde{X}_1-2}(\tilde{X}_1-1)\left(\frac{\tilde{X}_1}{n} - \frac{(\tilde{X}_1)_2}{(n)_2}\right)\right)$$

and

$$q_{3,2} = \mathbf{E}\mathbf{P}_X\{d_1(v) = 1, d_1(u) = 0, \{u,v\} \subset \mathcal{V}_1\}$$
$$= \mathbf{E}(\mathbf{P}_X\{d_1(v) = 1, d_1(u) = 0|\{u,v\} \subset \mathcal{V}_1\}\mathbf{P}_X\{\{u,v\} \subset \mathcal{V}_1\})$$
$$= \mathbf{E}\left(Q_1(1-Q_1)^{2\tilde{X}_1-4}(\tilde{X}_1-2)\frac{(\tilde{X}_1)_2}{(n)_2}\right).$$

We obtain

$$q_3 = \mathbf{E}\left(\frac{(\tilde{X}_1)_2}{n}Q_1(1-Q_1)^{\tilde{X}_1-2}\right) - \mathbf{E}\left(\frac{(\tilde{X}_1)_2}{(n)_2}\theta\right) = \frac{\tau}{n} - \mathbf{E}\left(\frac{(\tilde{X}_1)_2}{(n)_2}\theta\right),$$

where $0 \le \theta \le 1$ stands for the difference of two probabilities

$$\begin{aligned}\theta &= Q_1(1-Q_1)^{\tilde{X}_1-2}(\tilde{X}_1-1) - Q_1(1-Q_1)^{2\tilde{X}_1-4}(\tilde{X}_1-2)\\ &= \mathbf{P}_X\{d_1(v)=1\} - \mathbf{P}_X\{d_1(v)=1, d_1(u)=0\}\\ &\le \mathbf{P}_X\{d_1(v)=1\} = Q_1(1-Q_1)^{\tilde{X}_1-2}(\tilde{X}_1-1).\end{aligned}$$

□

Now we are ready to prove Lemma 1.

Proof of Lemma 1. For $\lambda_{n,m,k} \to +\infty$ Lemmas 2 and 3 imply $\mathbf{E}N'_{k-1} \to +\infty$ and

$$N_{k-1} = (1+o_P(1))N'_{k-1} = (1+o_P(1))\mathbf{E}N'_{k-1}$$

Hence $\mathbf{P}\{N_{k-1} \ge 1\} \to 1$.

Assume now that $\lambda_{n,m,k} \to -\infty$. Then $\lambda_{n,m,t} \to -\infty$ for $t = 1, \ldots, k$. For $h = 1, \ldots, k-1$ relations (9), (10) of Lemma 2 imply that

$$\mathbf{E}N'_h = o(1), \qquad \mathbf{E}|N_h - N'_h| = o(1).$$

We have $|\mathbf{E}N_h| \le \mathbf{E}|N_h - N'_h| + |\mathbf{E}N'_h| = o(1)$ and hence $N_h = o_P(1)$.

For $h = 0$ the bound $N_h = o_P(1)$ follows from the fact that $\lambda_{n,m,1} \to -\infty$ implies that $G_{[n,m]}$ is connected with high probability, see [1]. □

Acknowledgments. We thank referees for valuable references to related work and for useful comments.

Disclosure of Interests. The authors have no competing interests to declare.

References

1. Ardickas, D., Bloznelis, M.: Connectivity threshold for superpositions of Bernoulli random graphs. arxiv:2306.08113v2 (2023)
2. Bergman, E., Leskelä, L.: Connectivity of random hypergraphs with a given hyperedge size distribution. Discret. Appl. Math. **357**(15), 1–13 (2024)
3. Bloznelis, M., Karjalainen, J., Leskelä, L.: Normal and stable approximation to subgraph counts in superpositions of Bernoulli random graphs. J. Appl. Probab. **61**, 401–419 (2024)
4. Bloznelis, M., Leskelä, L.: Clustering and percolation on superpositions of Bernoulli random graphs. Random Struct. Algorithms **63**(2), 283–342 (2023)
5. Bloznelis, M., Marma, D., Vaicekauskas, R.: Connectivity threshold for superpositions of Bernoulli random graphs. II. arXiv:2311.09317 (2023)

6. Erdős, P., Rényi, A.: On random graphs. I. Publ. Math. Debrecen **6**, 290–297 (1959)
7. Frieze, A., Karoński, M.: Introduction to Random Graphs. Cambridge University Press, Cambridge (2015)
8. Godehardt, E., Jaworski, J., Rybarczyk, K.: Random intersection graphs and classification. In: Decker, R., Lenz, H.J. (eds.) Advances in Data Analysis, pp. 67–74. Springer, Heidelberg (2007)
9. Gröhn, T., Karjalainen, J., Leskelä, L.: Clique and cycle frequencies in a sparse random graph model with overlapping communities. Stohastic models (2024)
10. van der Hofstad, R.: Random Graphs and Complex Networks. Cambridge Series in Statistical and Probabilistic Mathematics, vol. 1. Cambridge University Press, Cambridge (2017)
11. van der Hofstad, R., Komjáthy, J., Vadon, V.: Random intersection graphs with communities. Adv. Appl. Probab. **53**, 1061–1089 (2021)
12. Janson, S., Łuczak, T., Ruciński, A.: Random Graphs. Wiley, New York (2000)
13. Petti, S., Vempala, S.S.: Approximating sparse graphs: the random overlapping communities model. Random Struct. Algorithms **61**, 844–908 (2022)
14. Yang, J., Leskovec, J.: Community-affiliation graph model for overlapping network community detection. In: IEEE 12th International Conference on Data Mining, pp. 1170–1175. IEEE (2012)
15. Yang, J., Leskovec, J.: Structure and overlaps of ground-truth communities in networks. ACM Trans. Intell. Syst. Technol. **5**(2), 1–35 (2014)

Improving Community Detection via Community Association Strength Scores

Jordan Barrett[1(✉)], Ryan DeWolfe[1], Bogumił Kamiński[2], Paweł Prałat[1], Aaron Smith[3], and François Théberge[4]

[1] Department of Mathematics, Toronto Metropolitan University, Toronto, Canada
{jordan.barrett,ryan.dewolfe,pralat}@torontomu.ca
[2] Decision Analysis and Support Unit, SGH Warsaw School of Economics, Warsaw, Poland
bkamins@sgh.waw.pl
[3] Department of Mathematics and Statistics, University of Ottawa, Ottawa, Canada
asmi28@uOttawa.ca
[4] Tutte Institute for Mathematics and Computing, Ottawa, Canada
theberge@ieee.org

Abstract. Community detection methods play a central role in understanding complex networks by revealing highly connected subsets of entities. However, most community detection algorithms generate partitions of the nodes, thus (i) forcing every node to be part of a community and (ii) ignoring the possibility that some nodes may be part of multiple communities. In our work, we investigate three simple community association strength (CAS) scores and their usefulness as post-processing tools given some partition of the nodes. We show that these measures can be used to improve node partitions, detect outlier nodes (not part of any community), and help find nodes with multiple community memberships.

1 Introduction

Detecting and analyzing the community structure in graphs is a fundamental problem in applied graph theory. Algorithms such as Girvan-Newman [22], Louvain [2], and Leiden [27] attempt to partition the nodes of a graph into clusters with the general goal of maximizing edge density within clusters and minimizing edge density between clusters. These global partitioning algorithms have been used to detect community structure in various real-world networks such as collaboration networks [20,22], biological networks [3,26], and social media networks [9,23], and have been tested and studied on synthetic models with ground-truth communities such as ABCD [11] and LFR [18]. Typically, these community detection algorithms return a partition, meaning each node must be part of exactly one community. In many real-world networks, though, members can belong to 0, 1, or multiple communities. Some algorithms to detect overlapping communities include clique-percolation [5], edge clustering [1] and ego-split [6].

A post-processing algorithm was proposed [10] to detect overlapping communities, but this approach is based on k-core clustering and often only covers a small part of the nodes.

In this paper, we use local community aware scores called *community association strength* (CAS) scores as a post processing tool for improving community detection in networks. These CAS scores are inspired by the measures proposed in [12]. The rest of the paper is organized as follows. In Sect. 2, we outline three CAS scores and show their similarities, differences, and relative abilities to predict community involvement. In Sect. 3, we show how such CAS scores can be used for three different but complementary tasks: (i) improving an existing graph partitioning algorithm, (ii) detecting nodes not part of any community (outlier nodes), and (iii) detect multi-community node memberships. Finally, some concluding remarks are given in Sect. 4

2 Community Association Strength

2.1 CAS Scores

Given a graph $G = (V, E)$, a CAS score is any function $f : V \times 2^V \to [0, 1]$. In practice, we aim to find a CAS score f such that $f(v, C)$ indicates how well v is associated to community C, i.e., $f(v, C) \approx 1$ should imply that v is strongly associated to community C whereas $f(v, C) \approx 0$ should imply that v has little to no association to community C. We will motivate and outline three such functions.

Internal Edge Fraction (IEF). Arguably, the simplest measure for how strongly v is associated with community C is the *internal edge fraction*. For a community $C \subseteq V$, write $\deg_C(v) := |\{(u,v) \in E : u \in C\}|$ and $\deg(v) := \deg_V(v)$. Then, the internal edge fraction of v in C is defined as

$$\text{IEF}(v, C) := \frac{\deg_C(v)}{\deg(v)}.$$

Although the internal edge fraction is both easy to interpret and a strong indicator of community association, it fails to account for the size of the communities. For example, if G contains a large community C_1 with, say, $|C_1| \approx |V|/2$ and a small community C_2 with $|C_2| \ll |V|$, then we may want to distinguish the outcomes $\text{IEF}(v, C_1) = 1$ and $\text{IEF}(v, C_2) = 1$, especially if $\deg(v)$ is large.

Normalized Internal Edge Fraction (NIEF). For a graph $G = (V, E)$ and a community C, write $w(C) := \text{vol}(C)/\text{vol}(V)$ where $\text{vol}(C) := \sum_{v \in C} \deg(v)$. The second CAS score we consider is the *normalized internal edge fraction*, defined as

$$\text{NIEF}(v, C) := \max\{\text{IEF}(v, C) - w(C), 0\}.$$

This CAS score is derived as follows. Let $G = (V, E)$, let $\mathbf{d} = \{\deg(v), v \in V\}$ and let $\widehat{G} \sim \text{ChungLu}(\mathbf{d})$ [4]. Then, for any $v \in V$,

$$\text{NIEF}(v, C) := \max\{\text{IEF}_G(v, C) - \mathbb{E}[\text{IEF}_{\widehat{G}}(v, C)], 0\},$$

where we assume that C is fixed before sampling $\widehat{G}$. In other words, NIEF(v, C) compares the actual internal edge fraction of v to the "expected" such fraction under the associated null-model. Note that this score is inspired by the *community association strength* feature presented in [12].

Consider again the previous example with G containing communities C_1 and C_2 with $|C_1| \approx |V|/2$ and $|C_2| \ll |V|$. If IEF$(v, C_1) = 1$ and IEF$(v, C_2) = 1$ then we get that NIEF$(v, C_1) \approx 0.5$ and NIEF$(v, C_2) \approx 1$. These scores agree with the intuition that v having many connections into a small community is more surprising than having the same number of connections into a large community.

P Score. Finally, we consider a score based on the classic *p*-value significance test. Write $F(\cdot; n, p)$ for the CDF of the distribution Binomial(n, p). The P score is defined as
$$\mathrm{P}(v, C) := F\left(\deg_C(v) - 1; \deg(v),\ \mathrm{w}(C)\right).$$
Here, $1 - \mathrm{P}(v, C)$ is the probability that at least $\deg_C(v)$ edges join v and C in a Chung-Lu resampling of G (keeping C fixed), i.e., $1 - \mathrm{P}(v, C)$ is the classic *p*-value of $\deg_C(v)$ when considering the Chung-Lu model as the null model. Similar to the NIEF score, the P score prefers smaller communities. We will explore the similarities and differences between the three measures in the next section. Note that the P score is similar to, albeit distinct from, the community fitness measure used in [19].

These three measures are by no means a comprehensive list of CAS scores. Rather, we have chosen these scores to show how CAS can be used to analyse graphs with community structure containing outliers and overlapping communities.

2.2 Comparing CAS scores

All three CAS scores, IEF, NIEF, and P, share two qualitative properties that we believe are good indicators of the association strength of a node into a community. Firstly, all three scores evaluate 0 if no edges join v and community C. Secondly, for fixed $\mathrm{vol}(C)$, all three scores are monotone with respect to $\deg_C(v)$; if $\mathrm{vol}(C_1) = \mathrm{vol}(C_2)$ and $\deg_{C_1}(v) > \deg_{C_2}(v)$ then (v, C_1) yields a higher CAS score than (v, C_2) under all three measures.

Perhaps more interesting than the similarities between these three CAS scores are their differences. A key difference is how the measures change based on the community size and the degree of the node. We will highlight this difference by comparing the measures under two different transforms. Given a graph $G = (V, E)$ and a community C, consider the following transforms.

1. Add a disconnected k-clique to the graph and assign each new vertex to community C.
2. Create a second copy of each edge in E.

Now let G be a graph, C be a community in G, v be a vertex in C, and let G_1 and G_2 be constructed from G by the respective transforms. Firstly, we have that $\text{IEF}_{G_1}(v, C) = \text{IEF}_{G_2}(v, C) = \text{IEF}_G(v, C)$, as the fraction of edges into C has not changed under either transform. Next, note that $\mathbb{E}[\text{IEF}_{G_1}(v, C)] > \mathbb{E}[\text{IEF}_G(v, C)]$ whereas $\mathbb{E}[\text{IEF}_{G_2}(v, C)] = \mathbb{E}[\text{IEF}_G(v, C)]$, implying that $\text{NIEF}_{G_1}(v, C) < \text{NIEF}_G(v, C)$ and $\text{NIEF}_{G_2}(v, C) = \text{NIEF}_G(v, C)$. Lastly for the P score, we have that $P(v, G_1) < P(v, G)$, but the relationship between $P(v, G)$ and $P(v, G_2)$ is more complicated. Indeed, if $P(v, G) = 1 - \epsilon$ for $\epsilon \ll 1$ then $P(v, G_2) > P(v, G)$, whereas if $P(v, G) = \epsilon \ll 1$ then $P(v, G_2) < P(v, G)$, and this is a consequence of the law of large numbers. To summarize, $\text{IEF}(v, C)$ is invariant under transform 1 which increases $\text{vol}(C)$ and under transform 2 which doubles the number of edges, $\text{NIEF}(v, C)$ changes under transform 1 but not 2, and $P(v, C)$ changes under both transforms.

Let us continue with two examples showing this result in action.

Example 1. Let $G = (V, E)$ be a graph with $\text{vol}(V) = 10{,}000$, let $C_1, C_2 \subset V$ be communities in G with $\text{vol}(C_1) = \text{vol}(V)/2 = 5{,}000$ and $\text{vol}(C_2) = 100$, and let $v \in V$ be a node with $\deg(v) = 5$ such that all edges from v connect to either C_1 or C_2. Table 1 shows the three CAS scores as $\deg_{C_1}(v)$ and $\deg_{C_2}(v)$ vary, highlighting the point in each score where the association strength of v into C_2 becomes larger than that of C_1.

We find that $\text{IEF}(v, C_1) < \text{IEF}(v, C_2)$ precisely when $\deg_{C_1}(v) < \deg_{C_2}(v)$. However, for both NIEF and P, there is a large penalty associated with C_1 compared to a negligible penalty associated with C_2. Note that in this case, the bias towards the smaller community is stronger for P than for NIEF. In our testing, we found this trend to be true in general.

Table 1. A comparison of CAS scores for 2 communities $C_1, C_2 \subset V$ with $\text{vol}(V) = 10{,}000$, $\text{vol}(C_1) = 5{,}000$ and $\text{vol}(C_2) = 100$. The scores are rounded to 2 decimal places, and node v is omitted from the notation. Here, $\deg(v) = 5$ and each row represents a different split of $\deg(v)$ into C_1 and C_2. The grey cells highlight when a CAS score favours C_2 over C_1.

$\deg_{C_1}$	$\deg_{C_2}$	$\text{IEF}(C_1)$	$\text{IEF}(C_2)$	$\text{NIEF}(C_1)$	$\text{NIEF}(C_2)$	$P(C_1)$	$P(C_2)$
5	0	1	0	0.5	0	0.97	0
4	1	0.8	0.2	0.3	0.19	0.81	0.95
3	2	0.6	0.4	0.1	0.39	0.5	1
2	3	0.4	0.6	0	0.59	0.19	1
1	4	0.2	0.8	0	0.79	0.03	1
0	5	0.0	1.0	0	0.99	0	1

Example 2. Consider a graph $G = (V, E)$, a community C with $\text{vol}(C) = \text{vol}(V)/2$, and a node $v \in V$. Table 2 shows NIEF and P scores for (v, C) as $\deg(v)$ and $\deg_C(v)$ vary whilst $\deg_C(v)/\deg(v)$ remains constant.

The increasing values of P(v, C) as deg(v) and $\deg_C(v)$ increase is a consequence of the law of large numbers for the binomial distribution. Ultimately, what this example shows is that the P score, conditioned on $\deg_C(v)/\deg(v) > w(C)$, is biased towards nodes with a larger degree, whereas the NIEF score has no such bias.

Table 2. The comparison of 2 CAS scores on community C with $\text{vol}(C) = \text{vol}(V)/2$. The scores are rounded to 2 decimal places, and the node v is omitted from the notation.

deg	$\deg_C$	NIEF(C)	P(C)
3	2	0.17	0.5
6	4	0.17	0.66
9	6	0.17	0.75
12	8	0.17	0.81
15	10	0.17	0.85

2.3 The Quality of CAS Scores

To test the quality of the various community association strength scores, we need graphs with ground-truth communities, where the communities do not necessarily form a partition of the nodes. To this end, we will test a family of synthetic models stemming from the Artificial Benchmark for Community Detection (ABCD) model. The ABCD model [13] is a synthetic model with ground-truth communities and has 8 parameters to control the number of nodes, the degree and community size distributions, and the fraction ξ of noise; $1 - \xi$ is (roughly) the fraction of edges of a node into its community. A fast, multi-threaded implementation of ABCD (ABCDe) was introduced in [17], and a hypergraph generalization (h-ABCD) was introduced in [16]. Importantly for our research, the ABCD model was generalized (ABCD+o) to include outliers [15], and further generalized (ABCD+o^2) to allow for overlapping communities. The latter model includes a parameter $\eta \geq 1$ which governs the expected number of community memberships for non-outlier nodes.

We first test the CAS scores' ability to rank communities based on the likelihood of a node being in those communities. For a graph G, a collection of ground-truth communities $\mathcal{C}$, and a CAS score $f : V \times 2^V \to [0, 1]$, let $V_k \subseteq V$ be the set of nodes such that $v \in V_k$ if and only if v is contained in at least k communities in $\mathcal{C}$. For each score f, each relevant k, and each $v \in V_k$, we consider the kth highest ranking community in $\mathcal{C}$ according to f and check if this community indeed contains v. Figure 1 presents the experiment results using ABCD+o^2 graphs with two different noise parameters: $\xi = 0.35$ and $\xi = 0.65$. All three measures perform similarly, with NIEF and P performing slightly better. The results suggest that each of the measures can accurately predict 1 or 2 communities a node is a member of, and with a low noise parameter, the prediction accuracy remains high as the number of communities increases.

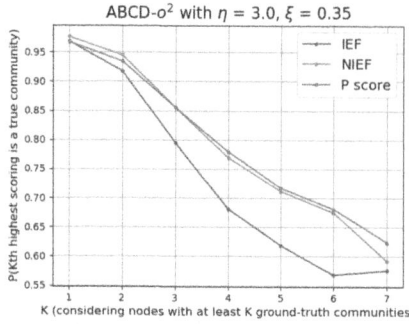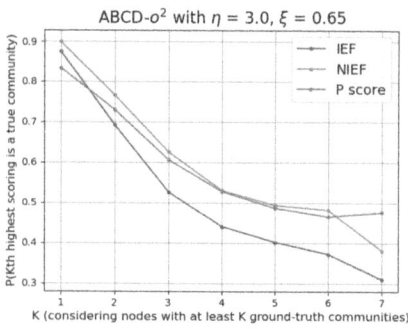

Fig. 1. Proportion of Kth highest scoring community that are actually ground-truth communities for the three CAS scores. Results are averaged over 10 ABCD+o^2 graphs with 10,000 nodes including 250 outliers, overlap parameter $\eta = 3$ and noise parameters $\xi = 0.35$ (left) and $\xi = 0.65$ (right). We can see that with all scores, the first few highest-scoring communities are almost always ground-truth, and this slowly degrades as K increases, with the NIEF and P scores decreasing more slowly.

Next, we test the scores' ability to distinguish outliers from non-outliers. Let $V_o \subseteq V$ be the outlier nodes concerning ground-truth communities $\mathcal{C}$. For each CAS score f, we order the nodes in V from smallest to largest based on $\max\{f(v,\cdot)\}$. We predict that, for $u,v \in V$ with $\max\{f(u,\cdot)\} < \max\{f(v,\cdot)\}$, u is more likely to be an outlier than v. We then compare our prediction to the ground truth with a receiver operating characteristic (ROC) curve. Figure 2 summarizes the results of this experiment by showing ROC curves for ABCD+o^2 graphs with 10,000 nodes (including 250 outliers), moderate noise ($\xi = 0.55$) and two different values for η. We also show each score's area under the ROC curves (AUC). While all measures get almost perfect results when $\eta = 1$ (no community overlap), we see degradation in the presence of community overlap ($\eta = 3$). We again see that all three measures perform similarly, with NIEF performing slightly better.

These two experiments indicate that CAS scores can capture local properties useful to refine node partitions. In the next section, we consider more realistic experiments where we do not explicitly use the ground-truth communities (but we use them for evaluation).

3 Using CAS Scores in Practice

In the previous section, we tested the quality of CAS scores with respect to the ground-truth communities of a graph. In practice, however, we may not have access to the ground-truth communities and would thus like to leverage CAS scores to help recover said communities. To this end, we present three scenarios where CAS scores can be useful. Section 3.1 describes an improvement to an existing consensus clustering algorithm. We then show applications of CAS scores to post-process any partitioning algorithm: identifying outlier nodes in Sect. 3.2

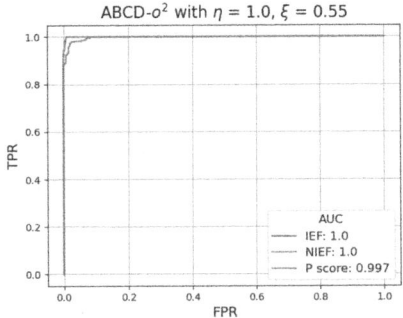 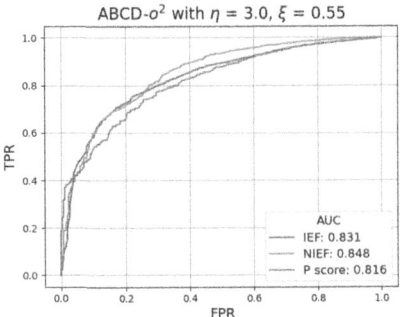

Fig. 2. ROC curves and corresponding AUC measures for three CAS scores for ABCD+o² graphs.

and identifying multi-community memberships in Sect. 3.3. Finally, we apply all three ideas to the well known college football graph in Sect. 3.4.

3.1 Improving Graph Partitions

In this section, we use CAS scores to modify an existing clustering algorithm called Ensemble Clustering for Graphs (ECG) [24,25] which is itself a modification of the Louvain algorithm. Let us briefly describe the ECG, and then our modification to ECG (which we will refer to as CAS-ECG).

The ECG algorithm takes a graph $G = (V, E)$ and a positive integer k as input and executes as follows.

1. Perform the first iteration of the standard Louvain algorithm on G, k times, to yield k "level 1" partitions of V.
2. Weight E such that the weight of an edge (u, v) is the fraction of partitions from step 1 with u and v in the same part.
3. Run Louvain, Leiden or some other partitioning algorithm on this weighted graph.

The full description of the algorithm can be found in Sect. 3 of [24].

We now detail CAS-ECG. During step 2 of ECG, edge (u, v) is weighted by how often u and v end up in the same part after 1 step of Louvain. Given some partition $\mathcal{C}$ of V with $u \in C_u$ and $v \in C_v$, write

$$\mathrm{ecg}\big((u,v),\mathcal{C}\big) = \begin{cases} 1 & C_u = C_v \\ 0 & C_u \neq C_v \end{cases}$$

Then the weight assigned to (u, v) at the end of step 2 of the ECG algorithm is $\sum \mathrm{ecg}\big((u,v),\mathcal{C}\big)$ where the sum is taken over the k partitions. We consider replacing this weight with a new weight based on a given CAS score. Intuitively, we want (u, v) to receive a higher weight if u is strongly associated with C_v and/or v is strongly associated with C_u, even when $C_u \neq C_v$. For a CAS scoring function f, we propose the following two options for weighting the edges.

$$f_{or}((u,v), \mathcal{C}) = f(u, C_v) + f(v, C_u) - f(u, C_v) \cdot f(v, C_u).$$
$$f_{and}((u,v), \mathcal{C}) = f(u, C_v) \cdot f(v, C_u).$$

We aim to find a CAS score f and a weighting scheme f_{or} or f_{and} that improves ECG. Thus, we test the performance of the six combinations of scores and weighting schemes to see if any can improve ECG's ability to recover ground-truth communities. We perform this test on ABCD graphs with 10,000 nodes, a minimum degree of 5, a minimum community size of 50, and varying levels of noise. Each clustering uses $k = 16$ runs in the ensemble step and obtains the final clustering from running the Leiden algorithm on the weighted graph. The results of this experiment are presented in Fig. 3. We find that each of the six modifications provides comparable results to the base ECG algorithm for $\xi < 0.55$. Although five out of the six configurations seem to yield comparable or worse results than ECG, we find that using P_{and} as the edge weighting function results in a substantial increase in the Adjusted Mutual Information (AMI) score between noise levels $\xi = 0.55$ and $\xi = 0.65$, with a peak increase of about 5%. As can be seen in the left plot, this range of noise levels corresponds to the "critical" region where improvements are significant; with lower noise values, most algorithms will yield good results, while for higher noise values, the resulting AMI scores are very low no matter which algorithm is used.

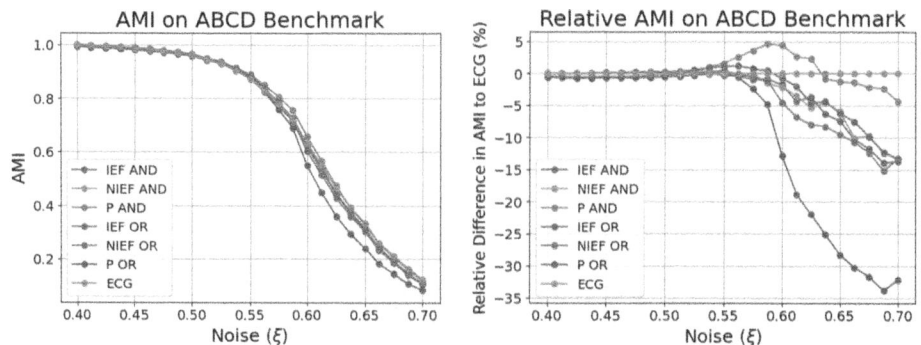

Fig. 3. (Left) Average AMI using each CAS and weight functions and the average relative change compared to the base ECG method (Right). Any option performs similarly to ECG, with P_{and} performing slightly better on graphs with ξ between 0.55 and 0.65. For each value of ξ, 50 ABCD graphs were generated.

3.2 Outlier Detection

This section uses a CAS score to post-process a partitioning algorithm to find outliers. Let $G = (V, E)$ be a graph with ground-truth communities and a set of outliers $V_o \subseteq V$, and let $\mathcal{C}$ be a partition of V found by a detection algorithm. Then, although $\mathcal{C}$ is not the ground truth assuming $|V_o| > 0$, we can still attempt to recover as much of V_o as possible by finding nodes v such that $CAS(v, C) \approx 0$ for all $C \in \mathcal{C}$. To test if this heuristic is feasible, we perform the following experiment on the ABCD+o model using each of the 3 CAS scores. First, we use the Leiden algorithm to obtain a partition of the nodes. We then use the CAS score of a node and its suggested community to rank the nodes from most likely to be an outlier (low maximum CAS score) to least likely to be an outlier (high maximum CAS score). Finally, we compute the area under the ROC curve (AUC) for our outlier prediction. For this experiment, we fix $|V| = 10,000$, and we fix the distributions for sampling degrees and community sizes. We test noise parameter ξ varying from 0.45 to 0.7, and a number of outliers varying from 100 to 4,000 (corresponding to %1 to %40 of the nodes). The results are shown in Fig. 4 and are averaged over 50 independent graphs for each configuration. We observe decreasing performance as the noise level ξ increases and slightly decreasing performance as the number of outliers increases. Each measure performs similarly, with a slight advantage for the NIEF score. These results show that on ABCD+o graphs and using Leiden to find an initial partition, all 3 CAS scores can fairly accurately recover the outliers when $\xi \leq 0.5$, whereas none of the scores can recover the outliers when $\xi \geq 0.6$.

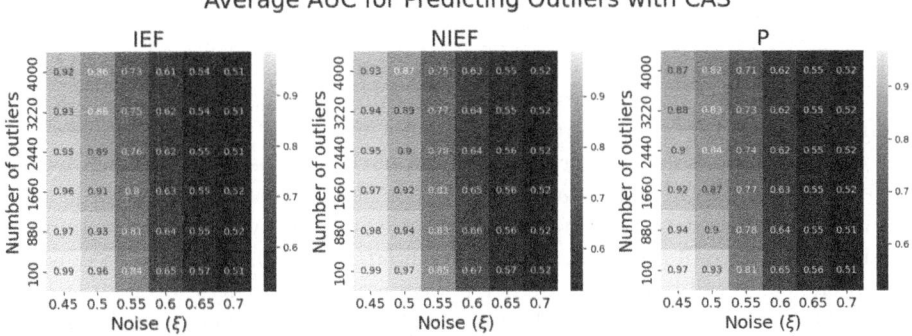

Fig. 4. Average AUC score using Leiden followed by CAS-based rankings on 50 ABCD-o graphs with 10,000 nodes.

3.3 Multi-communities

Finding overlapping communities in a graph is substantially more difficult than finding outliers, particularly when the graphs are noisy and/or the communities are numerous and small. It can quickly become difficult to distinguish noise edges from within-community edges. Here, we show how to use a CAS score to refine a collection of communities. Let $G = (V, E)$ be a graph, $\mathcal{C} = \{C_1, \ldots, C_k\}$ be a collection of communities in G found by some algorithm, f be a CAS score, and $\tau > 0$ be a threshold. Construct a new collection of communities, $\mathcal{C}'$, as follows.

1. Initially, $\mathcal{C}' = \{C'_1, \ldots, C'_k\}$ is a collection of empty sets.
2. For all $v \in V$ and all $C_i \in \mathcal{C}$, if $f(v, C_i) \geq \tau$ then add v to C'_i in $\mathcal{C}'$.

To test this refinement process, we use the NIEF score to refine communities obtained via the ego-split method. In this method, a partition is found on an auxiliary graph G'. This auxiliary graph contains one or more copy of each node from the original graph G, depending on the community structure of each node's ego-net. The partition is then mapped back to G, yielding overlapping communities (more details can be found in [6] and [14]). We consider ABCD+o[2] graphs with 10,000 nodes, a fixed noise level $\xi = 0.35$, and varying η values. The results are presented in Fig. 5. To measure the quality of predicted communities compared to the ground truth, we use the overlapping Normalized Mutual Information (oNMI) measure: a similarity measure for two collections of subsets $\mathcal{X}, \mathcal{Y}$ of a set S [21]. For each $\eta \in \{1, 1.5, 2, 2.5, 3\}$, we first find $\mathcal{C}_{guess}$ via the ego-split method (restricting the minimum community size to 10) and compute oNMI($\mathcal{C}_{guess}, \mathcal{C}_{true}$). We then obtain $\mathcal{C}'_{guess}$ using our refinement method with a variety of thresholds τ and compare the resulting oNMI scores with the original. We find that, while no single value for τ is clearly the best, choosing $\tau \in [0.075, 0.25]$ yields a refinement $\mathcal{C}'_{guess}$ that is better than the initial prediction $\mathcal{C}_{guess}$. This result suggests that, with a well-chosen CAS score and threshold, the refinement process can indeed improve existing detection algorithms.

In practical applications, since the ground truth is unknown, it is not always clear how to pick a good threshold for the refinement process. We propose a guiding method in Fig. 6 where we show the number of outliers obtained for each choice of threshold. The true number of outlier nodes is 250 (shown with a dashed line). While this information is also not likely known beforehand, it gives us some rule of thumb to set the threshold. For example, if we suspect the number of outliers to be small (as it is), then a threshold around 0.1 for NIEF seems like a good choice, possibly slightly higher if we suspect no or little overlap ($\eta \approx 1$) and slightly lower if we suspect lots of overlap. This threshold corresponds to good results in Fig. 5.

3.4 Illustration on a Real Graph

We now illustrate the methods described in Sects. 3.1, 3.2, and 3.3 using a real-world graph. We consider the college football graph from [8] with corrections to the labels as described in [7]. The graph has 115 nodes (teams) and 613 edges (games played). After the corrections, there are 12 communities corresponding to football conferences. In general, teams play most games within their conference.

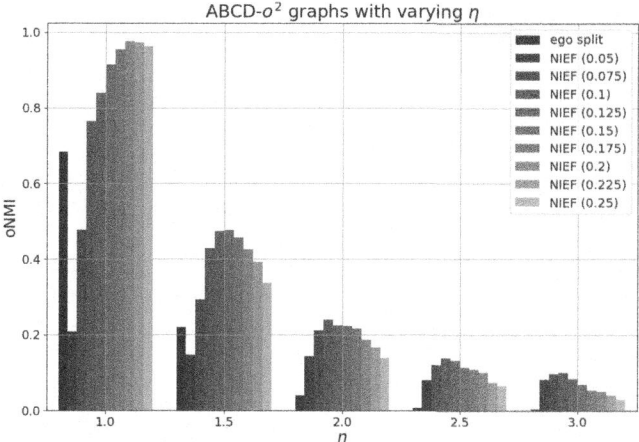

Fig. 5. Ego-splitting (darker bars) followed by NIEF with varying threshold values for ABCD+o^2 graphs with 10,000 nodes, $\xi = 0.35$, and varying η value. We compare each set of communities with the ground truth via the oNMI measure.

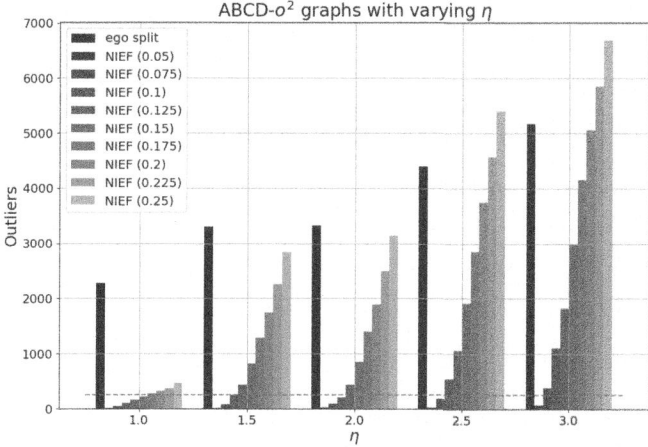

Fig. 6. Ego-splitting (darker bars) followed by NIEF with varying threshold values for ABCD+o^2 graphs with 10,000 nodes, $\xi = 0.35$, and varying η values. We compare the number of outlier nodes produced in each case.

One of these communities is in fact a group of independent teams which we use as a surrogate for outlier nodes.

We first consider the CAS-ECG algorithm from Sect. 3.1. In Fig. 7, we show the graph using a forced directed layout[1], where the node colours correspond to the different conferences, and the outlier nodes are shown as black triangles (left plot). Running the CAS-ECG algorithm using the P$_{and}$ weighting

---
[1] the Fruchterman-Reingold algorithm in Python-igraph.

scheme, the same force-directed layout algorithm yields the (much nicer) plot on the right. Moreover, we get slightly better communities with CAS-ECG than with Leiden, i.e. larger AMI values.

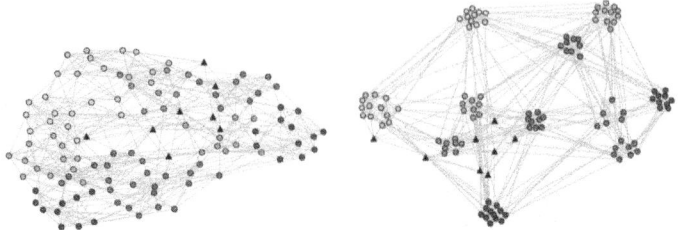

Fig. 7. Football graph displayed using a force-directed layout algorithm. The left plot has all unit edge weights, while the right plot has edge weights derived from the CAS-ECG algorithm. Communities are shown in colors, and outlier nodes as black triangles.

Next, we use the P score to rank nodes as possible outliers as per Sect. 3.2. The results are presented in Fig. 8. We run two versions of the experiment, the first using Leiden to obtain the initial partition and the second using the P_{and} version of CAS-ECG. We see that the eight outlier nodes are found in the top 10 (Leiden) or the top 8 (CAS-ECG). Moreover, the CAS-ECG algorithm slightly outperforms Leiden in recovering the ground-truth communities.

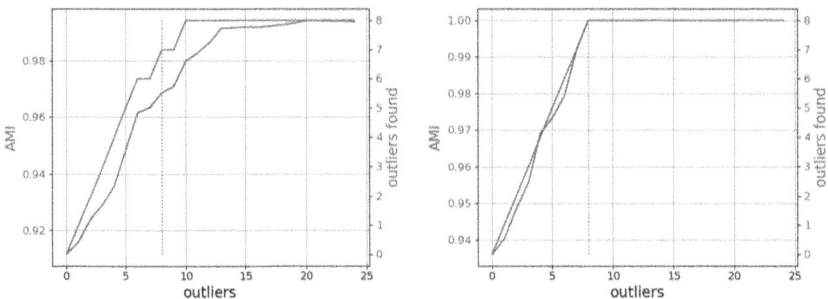

Fig. 8. Outlier detection on the football graph respectively using Leiden (left) and the P_{and} modified ECG (right), followed by node ranking via P score. We show the number of outliers found (blue curves) and the corresponding AMI (red curves) as we iterate through the ranked list. (Color figure online)

Finally, while there is no clear community overlap in this dataset, we look at the most likely node(s) that are part of multiple communities, again using the P_{and} version of CAS-ECG. In Fig. 9, we show the ego-nets for two nodes with high P scores for two communities. The nodes are shown as larger circles. In both cases, while the nodes are both part of a tight community (magenta), they seem to act as a bridge to other communities.

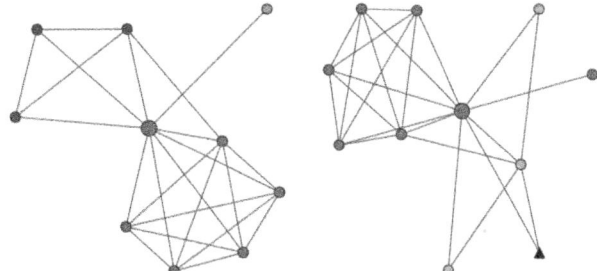

Fig. 9. Ego-nets for two nodes (shown with larger circles) having large CAS P-scores toward two communities.

4 Conclusion

We presented three community association strength functions, highlighted their similarities and differences, and showed their ability to recover community involvement in a network. We suggested multiple ways to leverage community association strength when detecting communities with overlap and outliers. While the experiments in Sect. 3 suggest that CAS scores do a good job at improving community detection, our goal is merely to show that improvement is *possible*. On the one hand, our experiments should be tested on a wide variety of real datasets with ground-truth communities to obtain more conclusive results. On the other hand, there might be stronger CAS scores and/or more clever uses of these scores that outperform what we have presented here. In an upcoming journal version, we will present more experiments on more datasets and delve deeper into the meaning behind the results. Furthermore, we leave it as an open problem to find CAS scores that can outperform the three scores presented here.

References

1. Ahn, Y.-Y., Bagrow, J.P., Lehmann, S.: Link communities reveal multiscale complexity in networks. Nature **466**(7307), 761–764 (2010)
2. Blondel, V.D., Guillaume, J.-L., Lambiotte, R., Lefebvre, E.: Fast unfolding of communities in large networks. J. Stat. Mech.: Theory Exp. **2008**(10), P10008 (2008)
3. Kuijjer, M.L., Calderer, G.: Community detection in large-scale bipartite biological networks. Front. Genet. **12** (2021). https://doi.org/10.3389/fgene.2021.649440
4. Graham, F.C., Lu, L.: Complex Graphs and Networks. CBMS Regional Conference Series in Mathematics, vol. 107. American Mathematical Society (2006)
5. Derényi, I., Palla, G., Vicsek, T.: Clique percolation in random networks. Phys. Rev. Lett. **94**(16), 160202 (2005)
6. Epasto, A., Lattanzi, S., Leme, R.P.: Ego-splitting framework: from non-overlapping to overlapping clusters. In: Proceedings of the 23rd ACM SIGKDD International Conference on Knowledge Discovery and Data Mining, pp. 145–154 (2017)

7. Evans, T.S.: Clique graphs and overlapping communities. J. Stat. Mech.: Theory Exp. **2010**(12), P12037 (2010)
8. Girvan, M., Newman, M.: Community structure in social and biological networks. Proc. Natl. Acad. Sci. **99**(12), 7821–7826 (2002)
9. Han, Z., et al.: H-Louvain: hierarchical Louvain-based community detection in social media data streams. Peer-to-Peer Netw. Appl. **17** (2024). https://doi.org/10.1007/s12083-024-01689-9
10. Jakatdar, A., Liu, B., Warnow, T., Chacko, G.: AOC: assembling overlapping communities. Quant. Sci. Stud. **3**(4), 1079–1096 (2022)
11. Kamiński, B., Pankratz, B., Prałat, P., Théberge, F.: Modularity of the ABCD random graph model with community structure. J. Complex Netw. **10**(6), cnac050 (2022). https://doi.org/10.1093/comnet/cnac050
12. Kamiński, B., Prałat, P., Théberge, F., Zajac, S.: Predicting properties of nodes via community-aware features. Soc. Netw. Anal. Min. **14**(1), 117 (2024). https://doi.org/10.1007/s13278-024-01281-2
13. Kamiński, B., Prałat, P., Théberge, F.: Artificial benchmark for community detection (ABCD)-fast random graph model with community structure. Netw. Sci. 1–26 (2021)
14. Kamiński, B., Prałat, P., Théberge, F.: Mining Complex Networks. Chapman and Hall/CRC (2021)
15. Kamiński, B., Prałat, P., Théberge, F.: Artificial benchmark for community detection with outliers (ABCD+O). Appl. Netw. Sci. **8**(1), 25 (2023)
16. Kamiński, B., Prałat, P., Théberge, F.: Hypergraph artificial benchmark for community detection (H-ABCD). J. Complex Netw. **11**(4), cnad028 (2023)
17. Kamiński, B., Olczak, T., Pankratz, B., Prałat, P., Théberge, F.: Properties and performance of the ABCDE random graph model with community structure. Big Data Res. **30**, 100348 (2022). https://doi.org/10.1016/j.bdr.2022.100348. https://www.sciencedirect.com/science/article/pii/S2214579622000429
18. Lancichinetti, A., Fortunato, S., Radicchi, F.: Benchmark graphs for testing community detection algorithms. Phys. Rev. E **78**(4), 046110 (2008)
19. Lancichinetti, A., Radicchi, F., Ramasco, J.J., Fortunato, S.: Finding statistically significant communities in networks. PLoS ONE **6**(4), e18961 (2011). https://doi.org/10.1371/journal.pone.0018961
20. Lužar, B., Levnajić, Z., Povh, J., Perc, M.: Community structure and the evolution of interdisciplinarity in Slovenia's scientific collaboration network. Plos One **9**(4), 1–5 (2014). https://doi.org/10.1371/journal.pone.0094429
21. McDaid, A.F., Greene, D., Hurley, N.: Normalized mutual information to evaluate overlapping community finding algorithms. arXiv preprint arXiv:1110.2515 (2011)
22. Newman, M., Girvan, M.: Finding and evaluating community structure in networks. Phys. Rev. E **69**(2), 026113 (2004)
23. Papadopoulos, S., Kompatsiaris, Y., Vakali, A., Spyridonos, P.: Community detection in social media. Data Min. Knowl. Discov. **24**, 515–554 (2012). https://api.semanticscholar.org/CorpusID:15719130
24. Poulin, V., Théberge, F.: Ensemble clustering for graphs. In: International Conference on Complex Networks and their Applications, pp. 231–243. Springer, Cham (2018)
25. Poulin, V., Théberge, F.: Ensemble clustering for graphs: comparisons and applications. Appl. Netw. Sci. **4**(1), 1–13 (2019). https://doi.org/10.1007/s41109-019-0162-z

26. Rahiminejad, S., Maurya, M.R., Subramaniam, S.: Topological and functional comparison of community detection algorithms in biological networks. BMC Bioinform. **20** (2019). https://doi.org/10.1186/s12859-019-2746-0
27. Traag, V.A., Waltman, L., Van Eck, N.J.: From Louvain to Leiden: guaranteeing well-connected communities. Sci. Rep. **9**(1), 5233 (2019)

PageRank Under Interpolation Between Undirected- and Directed Networks - A Case Study

Florian Henning[✉][iD], Remco van der Hofstad[iD], and Nelly Litvak[iD]

Eindhoven University of Technology, 5600MB Eindhoven, The Netherlands
{f.b.henning,r.w.v.d.hofstad,n.v.litvak}@tue.nl

Abstract. Among centrality measures, PageRank is particularly famous due to its implementation in the Google search engine. We have recently shown that in general undirected networks, the graph-normalized PageRank of any node in the network is bounded from above by its degree. This general statement, however, is not true in directed networks, where, e.g., the directed version of the preferential attachment model exhibits *heavier* tails for the limiting PageRank distribution than for the limiting in-degree-distribution. In this note, we illustrate the general upper bound on PageRank in undirected networks by scatter plots of datasets from three scale-free real-world networks of different sizes. We observe and explain a concentration phenomenon within the scatter plots. Furthermore, to shed light on how the directed-ness of edges changes the relation between degrees and PageRank, we interpolate between undirected and directed graphs as follows: for each of the three networks, we construct a new - directed - network by randomly choosing a subset of the edges of prescribed size and for each of its elements deleting exactly one of the two possible directions. As a result of this procedure, some small-degree vertices will obtain a PageRank that is above their in-degree.

We illustrate and explain this phenomenon for the chosen datasets by comparing to what happens in the configuration model.

Keywords: Undirected network · Network centrality · PageRank · Power-law hypothesis · Power-law distributions

1 Introduction

1.1 Scale-Free Networks

A characteristic property of many real-world networks, e.g., online social networks, the web graph or router networks, is the absence of a typical scale for the degrees of nodes within the network. This means that there is a small, yet highly relevant, proportion of nodes with very high degrees (e.g., [10]). On the level of the degree distribution $(p_k)_{k \in \mathbb{N}}$, the absence of a typical scale in such a network corresponds to a power-law degree distribution, i.e.,

$$p_k \approx ak^{-\tau},$$

for both the network size n and k large, where $\approx$ denotes an unspecified approximation. In particular, the probability for a uniformly chosen node from the network to have a degree larger than k (i.e., the tail distribution at k) is approximated by

$$\mathbb{P}(d_{V_n} > k) \approx bk^{-(\tau-1)},$$

for some $b > 0$. In this note, we investigate the relation between power-law degree distributions and the tails of the PageRank distribution. Let us next define PageRank, and more generally, centrality measures in networks.

1.2 Network Centrality Measures

Network centrality measures are employed to establish a hierarchy between nodes in a (potentially large) network, and to identify the influential ones. Presumably, the most simple and intuitive network centrality measures are the (in-)degree and Google's PageRank. In the recent thesis [15], comparison techniques for centrality measures (in particular the Closeness Centrality Curve (CCC) comparing the overlap of the k most relevant nodes according to the considered centrality measures) are discussed and applied to the established ones such as in- and out-degree, closeness, betweeness, harmonic centrality, Katz centrality and the PageRank with different damping factors. This comparison is done on real-word network data sets, as well as on directed and undirected versions of the configuration model.

In view of the results of [15], we focus the present work on the comparison between PageRank and the (in-)degree, based on what is known as the power-law hypothesis.

1.3 PageRank and the Power-Law Hypothesis

PageRank. PageRank [6] is an influential centrality measure initially introduced and implemented by the Google search engine. The PageRank distribution is nothing but the unique stationary distribution of the *easily-bored surfer* Markov chain, which, in each time step and with probability $c \in [0,1]$, takes a simple random walk step from its current position, and with probability $1 - c$ jumps to a node chosen uniformly from the entire network. Here, the *damping factor c interpolates* between being mainly local, or being more global instead. The PageRank vector is the solution to the linear system of equations

$$\pi(i) = c \sum_{j \to i} \frac{1}{d_j^{\text{out}}} \pi(j) + \frac{1-c}{n}.$$

To have PageRank on a similar scale as the (in-)degrees, it is convenient to go over to the graph-normalized PageRank, which is defined as

$$R(i) = n\pi(i),$$

where n is the graph size, so that the average PageRank R is equal to 1. Throughout this note, by PageRank we mean its graph-normalized version.

PageRank in Scale-Free Networks. Empirical data, particularly for the web graph, has shown that a power-law degree (in-)degree distribution often leads to a power-law for the PageRank distribution, with the same power-law exponent. The question of the generality of these empirical observations is coined the *power-law hypothesis*.

1.4 Heavier PageRank Tails in Directed Preferential Attachment Models

The (undirected) preferential attachment model (also called the *Barabási-Albert* model) [5] was introduced to describe the emergence of power laws in real-life networks. The preferential attachment model is a sequence of random graphs $(\text{PA}_n^{m,\delta})_{n \in \mathbb{N}}$, whose size linearly grows with the time n. There are several versions of the preferential attachment model, and we focus on one. In each time step $n+1$, a new vertex v_{n+1} having a fixed number m half-edges is added to the existing graph. Each of the m half-edges is, one after the other, connected to one of the existing vertices in the graph by a linear update rule (see [9, (1.3.65)] for specific instances of the update rule and also the introduction of [2] for further variations of the model):

$$\mathbb{P}(v_{j+1,n+1}^{(m)} \text{connects to } v_i^{(m)} \mid \text{PA}_n^{m,\delta}) = \frac{D_i(n,j) + \delta}{2m(n-1) + j + \delta n},$$

where $D_i(n,j)$ is the degree of vertex i after the jth edge of vertex v_{n+1} has been added and $\delta > -1$ is a fixed parameter. In the directed version of the preferential attachment model, we regard each of the m half-edges as an outgoing edge, i.e., all edges are directed from younger to older vertices in the network, all vertices have a fixed out-degree m, but can have (as n grows) an arbitrarily large in-degree.

Surprisingly, as shown in [4], its directed version violates the power-law hypothesis in that the PageRank distribution at a uniformly chosen vertex converges weakly (with respect to the size n of the graph tending to infinity) to a power law that has *heavier* tails than the weak limit of the in-degree distribution. On the other hand, this heavier PageRank tails phenomenon does not occur in the directed configuration model [7] nor, more generally, in the entire class of random inhomogeneous digraphs [11].

1.5 General Upper Bound in Undirected Networks

The above-mentioned heavier tails of the PageRank distribution in the directed preferential attachment model raises the question whether such heavier tails can also occur in undirected networks.

In [8] we proved that this is *not* possible. Indeed, in undirected graphs of any size (in particular, not only asymptotically) the PageRank of '*every node*' is bounded from above by the degree of the respective node. This result remains true (up to a scaling factor) in directed networks with a bounded ratio of in- and out-degrees.

For further results on the impact of this ratio in the context of the directed configuration model and inhomogeneous random digraphs, see, e.g., [14], and for results on personalized PageRanks on undirected networks with bounded ratio (w.h.p.) of maximal- and minimal degree, see [3].

2 Data Sets

We consider three different data sets that all describe undirected networks and are taken from the Stanford Large Networks Database. The largest of the three data sets is the *as-Skitter* [12] internet topology graph with 1,696,415 nodes and 11,095,298 edges. The second data set is the *musae-Twitch-de* [16] (**Mu**lti-scale **A**ttributed **N**ode **E**mbedding) friendship graph between German-language gamers on the Twitch platform consists of 9498 nodes and 153,138 edges. Finally, the *ego-Facebook* [13] data set on social circles, or friend lists, on Facebook consists of 4039 nodes and 88,234 edges.

3 Scatter Plots and Concentration Phenomena

To provide more insight into the relation between the graph-normalized PageRank and (in-)degrees beyond tail-distributions, we present our analysis by means of scatter plots shown in Fig. 1.

As a first visualization that Pagerank is upper bounded by degrees, in Fig. 1, we show scatter plots of the PageRank versus the degree for all vertices. We see that all points clearly lie under the identity line where the PageRank coincides with the degree.

Moreover, the scatter plots in Fig. 1 indicate that the pairs of degrees and PageRanks seem to concentrate around the graph of the linear function $x \mapsto x/\alpha$, where α is the average degree of the nodes. Moreover, the PageRanks are (for sufficiently high degrees) above the graph of the linear function $x \mapsto x/\beta$, where β is four times the average degree of the nodes. We will provide an explanation for these phenomena on the basis of the configuration model. While the networks that we analyze do not look like typical realizations of a configuration model (our networks generally have more triangles than in the configuration model), the configuration model is an established model in network analysis and our theoretical considerations below match to the concentration phenomena that become visible in the plots. Here, the *undirected configuration model* $CM_n(\boldsymbol{d})$ is a model for a random graph of size n which has a prescribed degree vector $\boldsymbol{d} \in \mathbb{N}_0^n$ as parameter. Each node i is assigned d_i *half-edges*. Let $|\boldsymbol{d}| = \sum_{i \in [n]} d_i$ be the total degree. Then, a random graph is sampled by choosing any permutation $\varphi \colon [|\boldsymbol{d}|] \equiv \{1, 2, \ldots, |\boldsymbol{d}|\} \to [|\boldsymbol{d}|]$ uniformly at random and afterwards connecting the jth half-edge to the $\varphi(j)$th half-edge, where j runs from 1 to $|\boldsymbol{d}|$, and we impose the condition that $\varphi(j) \neq j$ for all $j \in [|\boldsymbol{d}|]$.

Conditions on Degree Sequence. Let $D_n = d_{\phi_n}$ denote the degree of a vertex in G_n where ϕ_n is a uniformly chosen node from G_n. We impose the following two assumptions on the degree sequence:

- **Condition 1.** $D_n \stackrel{n\to\infty}{\longrightarrow} D$ in distribution; **and**
- **Condition 2.** $\mathbb{E}[D_n] \stackrel{n\to\infty}{\longrightarrow} \mathbb{E}[D]$.

Under these two conditions, the configuration model converges locally in probability towards the *unimodular branching process tree* [9], whose root-degree is distributed as D, and all other vertices have i.i.d. offspring-distribution $\tilde{D}$ according to the size-biased distribution minus one, i.e.,

$$\mathbb{P}(\tilde{D} = k) = \frac{(k+1)\mathbb{P}(D = k+1)}{\mathbb{E}[D]}. \tag{1}$$

The total degree (including the edge to the parent) is given by the sized-biased distribution $D^* = \tilde{D} + 1$, i.e.,

$$\mathbb{P}(D^* = k) = \frac{k\mathbb{P}(D = k)}{\mathbb{E}[D]}. \tag{2}$$

The following asymptotic lower bound on PageRank at a uniformly chosen node ϕ_n holds for the undirected configuration model:

Theorem 1 (Theorem 2.8 in [8]). *Consider the undirected configuration model $G_n = \mathrm{CM}_n(\boldsymbol{d}^{(G_n)})$ where the degree distribution satisfies Conditions 1 and 2. Let $c \in [0,1]$ be a constant (the damping factor for PageRank). Then the PageRank vector $\boldsymbol{R}^{(G_n)}$ satisfies that, for all n, k,*

$$\mathbb{P}(R^{(G_n)}_{\phi_n} > k) \leq \mathbb{P}(D_n > k), \tag{3}$$

while further, for any $\beta > \frac{4\mathbb{E}[D]}{c(1-c)}$,

$$\liminf_{k\to\infty} \liminf_{n\to\infty} \frac{\mathbb{P}(R^{(G_n)}_{\phi_n} > k/\beta)}{\mathbb{P}(D_n > k)} \geq 1. \tag{4}$$

While [8, Theorem 2.8] states that the PageRank *distribution* has thinner tails than the degree in undirected graphs, the result proved is in fact pointwise, in that the proof shows that in *any* undirected graph G, and all v in the vertex set of G,

$$R^{(G)}_v \leq d^{(G)}_v. \tag{5}$$

It is this relation that we will investigate empirically in this paper, and, in particular, how it is changed in graphs that become more directed using an interpolation scheme.

The proof of the lower bound (4) involves a first-order approximation of the PageRank. As the local limit of $\left(\mathrm{CM}_n(\boldsymbol{d}^{(G_n)})\right)_{n\in\mathbb{N}}$ is a branching-process tree with i.i.d.-offspring distribution for all vertices except the root we conjecture the following improvement with a potential proof being based on taking into account all terms of the series expansion and employing independence of degrees:

Conjecture 1. In the lower bound (4) in Theorem 1 we conjecture that the scaling factor β can be decreased by a factor $c(1-c)$. More explicitly, we conjecture that for every $\tilde{\beta} > 4\mathbb{E}[D]$ it holds

$$\liminf_{k \to \infty} \liminf_{n \to \infty} \frac{\mathbb{P}(R_{\phi_n}^{(G_n)} > k/\tilde{\beta})}{\mathbb{P}(D_n > k)} \geq 1. \tag{6}$$

We now want to get insight into the behavior of $\mathbb{E}\left[\frac{R_\phi}{D_\phi}\right]$ on the unimodular branching process tree with root ϕ. We can offer an interesting heuristic insight by assuming that the fractions $X(v) = \frac{R_v}{D_v}$ of the limiting PageRank and the respective degree along the vertices of the tree are independent of the rest of the degree sequence. While there is a subtle dependence in reality, this simplification helps us to understand how PageRank dissipates through the layers of the tree. We will use $\hat{R}$ to denote this independent version. We formalize the analysis in the next proposition. While this is an obvious simplification, to the best of our knowledge, these are the first calculations for the PageRank of an undirected configuration model.

Proposition 1. *Let $(D_\phi, (D_{ij})_{i,j \in \mathbb{N}})$ be an independent family of $\mathbb{N}$-valued random, where D_ϕ has distribution D and D_{ij}'s have distribution D^* that has a size-biased distribution of D as defined in (2). Let the random variables $(X_\phi, (X_{ij})_{i,j \in \mathbb{N}})$ satisfy the stochastic recursion*

$$X_\phi \stackrel{d}{=} (1-c)\frac{1}{D_\phi} + c\frac{1}{D_\phi}\sum_{j=1}^{D_\phi} X_{1,j}, \tag{7}$$

$$X_{i,j} \stackrel{d}{=} (1-c)\frac{1}{D_{i,j}} + c\frac{1}{D_{i,j}}\sum_{j=1}^{D_{i,j}-1} X_{i+1,j} + c\frac{1}{D_{i,j}}X_{i-1,1}, \quad i > 0. \tag{8}$$

For $i = \phi, 1, 2, \ldots$, denote $x_i = \mathbb{E}[X_{i,j}]$, which is equal for all $j = 1, 2, \ldots$. Then the column vector $\mathbf{x} = (x_\phi, x_1, x_2, \ldots)^T$ is defined by

$$\mathbf{x} = [I - cB]^{-1}\mathbf{b}. \tag{9}$$

Here, the matrix B in (9) is the probability transition matrix of a simple random walk on the non-negative integers reflected at $0 = \phi$; this random walk transitions from $i > 0$ to $i - 1$ with probability

$$b = \mathbb{E}\left[\frac{1}{D_{i,j}}\right] = \sum_{k=1}^{\infty}\frac{1}{k}\frac{k\mathbb{P}(D=k)}{\mathbb{E}[D]} = \frac{1}{\mathbb{E}[D]};$$

from $i > 0$ to $i + 1$ with probability $1 - b$, and from ϕ to 1 with probability 1. The vector $\mathbf{b}$ is given by $\mathbf{b} = (1-c) \cdot (a, b, b, \ldots)^T$, where $a = \mathbb{E}\left[\frac{1}{D}\right]$.

Proof. Using the tower rule (7), we get

$$x_\phi = \mathbb{E}\left[\mathbb{E}\left[(1-c)\frac{1}{D_\phi} + c\frac{1}{D_\phi}\sum_{j=1}^{D_\phi}X_{1,j}\Big|D_\phi\right]\right] = (1-c)a + cx_1. \qquad (10)$$

Similarly, from (8), we get

$$x_i = \mathbb{E}\left[\mathbb{E}\left[(1-c)\frac{1}{D_{i,j}} + c\frac{1}{D_{i,j}}\sum_{j=1}^{D_{i,j}-1}X_{i+1,j} + c\frac{1}{D_{i,j}}X_{i-1,1}\Big|D_{i,j}\right]\right]$$
$$= (1-c)b + c(1-b)x_{i+1} + cbx_{i-1}, \; i > 0. \qquad (11)$$

Denote $\mathbf{x} = (x_\phi, x_1, x_2, \ldots)^T$. Then system of linear equations (10), (11) can be written in the form

$$[I - cB]\mathbf{x} = \mathbf{b}.$$

Since $I - cB$ is invertible, this proves (9). □

Note that $a > b$, and this makes x_ϕ larger than x_i's. However, the difference between a and b is typically not that large. If we had $b = a$ then all the x_j's would be the same and equal to $b = 1/\mathbb{E}[D]$. This heuristic derivation intuitively explains the observed slope close to $1/\mathbb{E}[D]$ in the scatter plots.

4 Interpolation Between Undirectedness and Directedness in Real-World Networks

As discussed above, the behavior of PageRank on undirected networks can be quite different to that on directed networks. By means of a case study, we perform an oriented percolation very similar to [1]. Here, we at first identify the undirected network with a directed one with bi-directed edges and then uniformly choose a subset of prescribed size from the edges. This subset contains those edges which we will make single-directional. After sampling this subset, for any of its elements, we uniformly choose one of the two possible directions and delete it. Depending on the proportion p of edges that we make single-directional, the procedure will significantly alter the ratio of in- and out-degrees.

Interpolation Between Undirected and Directed Networks
Here, we first describe our randomized algorithm that interpolates between undirected and directed graphs:

Input: List of m undirected edges $E = \{\{i_1, j_1\}, \ldots \{i_m, j_m\}\}$, proportion $p \in [0, 1]$.

1. Create two lists E^- and E^+ of directed edges from LoE as follows:
   – For all $\{a, b\} \in E$ do $(a, b) \in E^-$ and $(b, a) \in E^+$.
2. Uniformly choose a subset $RaInd$ of $\{1, 2, \ldots, m\}$ of size $round(pm)$, where we recall that m denotes the total number of edges given as input for the algorithm.

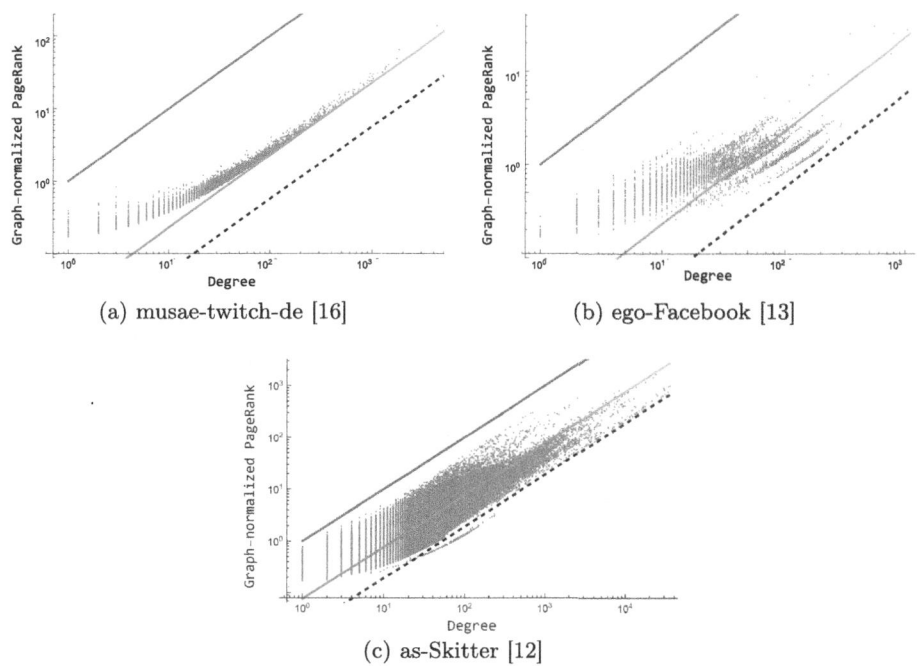

Fig. 1. The graph-normalized PageRank distribution with damping factor $c = 0.85$ (vertical axis) and the degrees (horizontal axis) with a $\log_{10} - \log_{10}$-scaling of the axes. All points lie below the red line corresponding to the graph of the identity function, which exemplifies the general upper bound for the PageRank in undirected networks. Furthermore, the point sets are approximately centered around the khaki-colored line that is vertically shifted by $-\log_{10} \alpha$ with α being the average degree of the respective graph. The black dashed line is vertically shifted by $-\log_{10} \tilde{\beta}$ with $\tilde{\beta}$ being four times the average degree as given in Conjecture 1.

3. For each $l \in RaInd$ uniformly choose among the two options: either delete the lth entry of the list E^-, or delete the lth entry of the list E^+ of reversed edges.
   Let $E'^- \subset E^-$ and $E'^+ \subset E^+$ denote the so-obtained lists of directed edges with deletions.
4. Generate a directed graph $G_{\text{dir}} = (V, E'^- \cup E'^+)$ from the lists with deletions. Here, V is the set of all vertices appearing in E.
   Generate graph $G_{\text{ref}} = (V, E^- \cup E^+)$ (the reference graph) from the list without deletions that contains each edge in both directions.

Note that in [1], each directed edge is removed with probability p, while in our case each edge can be removed with probability $p/2$, and we guarantee to keep an edge in at least one of the directions.

In Fig. 2 we see that as the proportion p of single-directed edges increases, some of the nodes will get a PageRank that is higher than the in-degree.

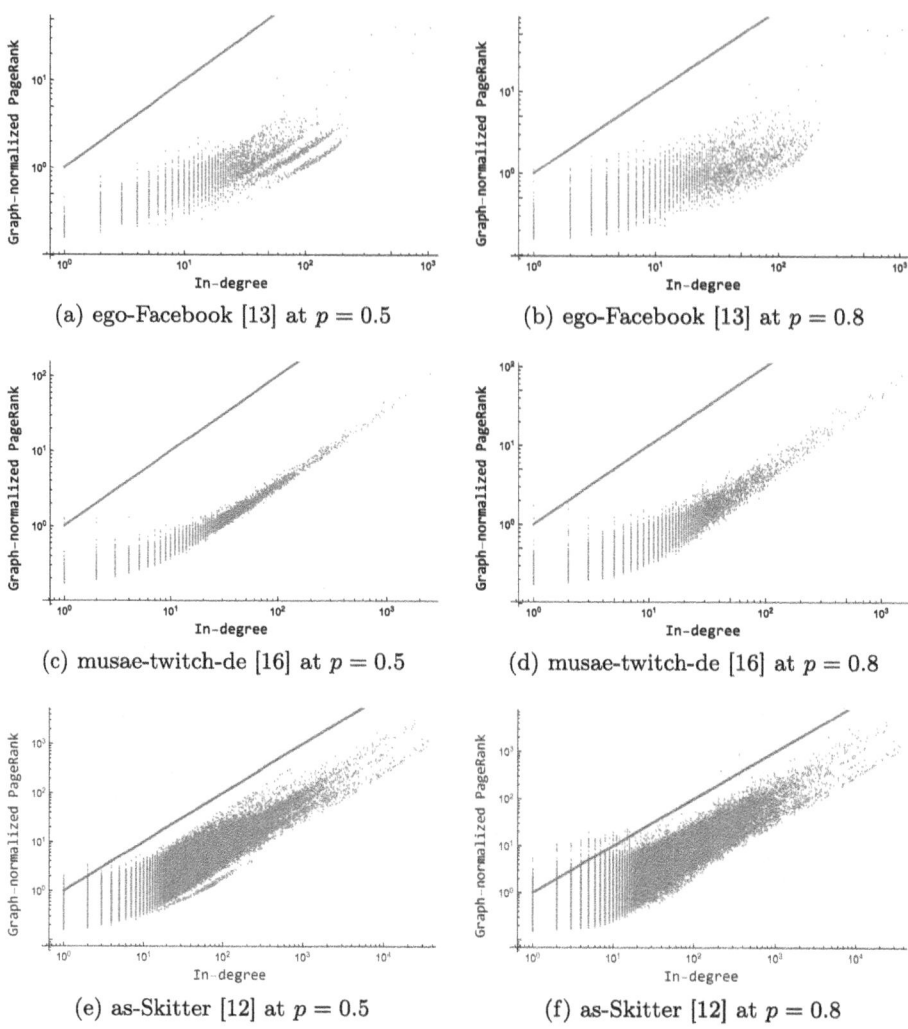

Fig. 2. Six realizations of the above algorithm applied to the three data sets and different values of the proportion p. The respective data plots for the initial undirected graphs are shown in Fig. 1. For each realization of the algorithm, the graph-normalized PageRank with damping factor $c = 0.85$ on the vertical axis are plotted against the in-degrees on the horizontal axis. As the proportion p of edges that have been made single-directional increases, some of the low-degree nodes get shifted above the red line, i.e., they obtain a graph-normalized PageRank that is higher than their in-degree.

The Parents of Nodes Whose PageRank Overshoots the In-Degree. To determine the main effects that contribute to this phenomenon, we further specifically look at those nodes in the truncated graph G_{dir} that have in-degree 1, but a PageRank that is larger than 1. Heuristically, from the definition of

PageRank, the unique parent of each of these nodes must have a large PageRank. For each of these nodes, we compare the in- and out-degree and PageRank of its unique ancestor with the respective values of these quantities for the (undirected) reference graphs.

For the two smaller datasets, musae-twitch-de and ego-Facebook, this reveals that the phenomenon of increasing PageRank seems to be mostly explained by a superposition of two effects on the parent of those nodes (see Tables 1a and 1b):

- The out-degree of the parent is reduced. Thus, the remaining children get a higher share of the total degree weight.
- The PageRank of the parent increases.

While the first effect might be not surprising with respect to the definition of and what is known about PageRank, the second effect might be more surprising. We conjecture that the second effect is explained by the first effect affecting the nodes which remain a directed edge pointing towards the previously considered node (the parent's parents).

A Positive Proportion of Vertices with In-Degree 1 and Large PageRank. We next explain, based on the first of the two effects described above, how a positive proportion of vertices with in-degree 1 and large PageRank can arise.

Proposition 2. *Consider any undirected graph that has a local limit with a positive proportion $q_{\ell,k}$ of vertices that have degree equal to some fixed small ℓ (for example $\ell = 1$), and that have a neighbor of degree k for k large. Then, for k sufficiently large the following holds true:*
with a positive probability, the graph obtained from applying the randomized algorithm that makes a proportion p of edges single-directed will contain at least one vertex with a PageRank larger than the in-degree.

Proof. Consider a pair (v_l, w_k) of neighboring vertices in the local limit such that v_l has degree ℓ and w_k has degree k. After applying the above randomized algorithm to make a proportion p of edges directed, the probability that vertex v_ℓ will have in-degree 1, getting the directed edge from w_k, equals $p^\ell \cdot (1/2)^\ell$, while the probability that w_k directs all its remaining $k-1$ edges which are not attached to v_ℓ to in-edges for itself is $p^{k-1} \cdot (1/2)^{k-1}$. This implies that the degree-k neighbor w_k sends all of its PageRank to the vertex of interest. Since $R_u \geq 1 - c$ for every vertex u, this means that the PageRank of the vertex v_ℓ will be at least $kc(1-c)$, which can be made arbitrarily large by making k large. In particular, we can make this PageRank larger than 1. This shows that with a positive probability, after directing the edges, there will be a positive (though possibly quite small) proportion of vertices with degree 1 and PageRank larger than 1. □

Table 1. Parameters of the unique parents of those nodes in Figs. 2b and 2d that get shifted above the line and have in-degree 1. The node numbers correspond to the ones used in the data sources.

(a) In Figure 2b, nodes 3984, 4010, 4024, 4035 and 4036 get shifted above the line and have in-degree one. They all have the common unique parent 3980.

Data for unique parent 3980	before removal	after removal
In-degree	59	58
Out-degree	59	11
PageRank	8.71032	12.2927

(b) In Figure 2d, nodes 5918, 6150 and 1462 get shifted above the line and have in-degree one. The corresponding parents are 18, 967 and 8114

Data for the parents	before removal	after removal
In-degrees	2; 16; 3	2; 16; 3
Out-degrees	2; 16; 3	1; 1; 1
PageRanks	0.470515; 0.588953; 0.433619	1.83977; 1.00621; 1.21459

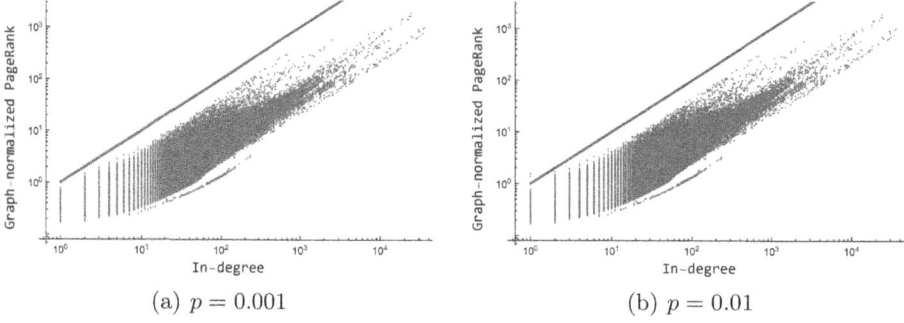

(a) $p = 0.001$ (b) $p = 0.01$

Fig. 3. Two realizations of the above algorithm for smaller values of p and the as-Skitter [12] dataset, again with damping factor $c = 0.85$. We see that already at $p = 0.01$ some small-degree nodes obtain a higher PageRank than their in-degree.

Large PageRanks at Low Proportion p of Single-Directed Edges. In the two smaller networks, a large value (0.5 or higher) for the proportion p of single-directed edges is needed to evoke the effect of PageRanks becoming larger than the corresponding in-degrees (cf. Figs. 2a to 2d).

5 Outlook and Discussion

In this paper, we have investigated how PageRank can become larger than the in-degree in directed networks, while PageRank is bounded by the degree in undirected networks. We have done this by interpolating between the undirected version of the graphs, and the randomly truncated directed versions of the real-world networks. We have studied which nodes are such that the PageRank is larger than the in-degree. This occurs for vertices that have low in-degree, and have a unique high-degree parent for whom relatively many edges have been changed to in-edges.

Our work raises the following follow-up questions. It remains to formally prove the concentration results that become visible in Fig. 1 on the level of random graph models.

6 Data Availability

The plots shown in Figs. 1, 2 and 3 have been generated using Wolfram Mathematica and using the datasets mentioned in the respective captions. The underlying realizations of the above randomized algorithm presented in Figs. 2 and 3 and Tables 1b and 1a have been obtained with Wolfram Mathematica. The lists of edges for the resulting directed graphs have been saved as .csv-files and are available upon request.

Acknowledgments. The authors thank the three anonymous referees for their insightful comments on the first version of this paper.

The work of RvdH and NL is supported by the Netherlands Organisation for Scientific Research (NWO) through the Gravitation NETWORKS grant 024.002.003, and by the National Science Foundation under Grant No. DMS-1928930 while the authors were in residence at the Simons Laufer Mathematical Sciences Institute in Berkeley, California, during the Spring semester.

Disclosure of Interests. The authors have no competing interests to declare that are relevant to the content of this article.

References

1. Alimohammadi, Y., Borgs, C., Saberi, A.: Locality of random digraphs on expanders. Ann. Probab. **51**(4), 1249–1297 (2023). https://doi.org/10.1214/22-AOP1618
2. Antunes, N., Banerjee, S., Bhamidi, S., Pipiras, V.: Attribute network models, stochastic approximation, and network sampling and ranking algorithms. Preprint (2023)
3. Avrachenkov, K., Kadavankandy, A., Ostroumova Prokhorenkova, L., Raigorodskii, A.: PageRank in undirected random graphs. In: Gleich, D.F., Komjáthy, J., Litvak, N. (eds.) WAW 2015. LNCS, vol. 9479, pp. 151–163. Springer, Cham (2015). https://doi.org/10.1007/978-3-319-26784-5_12
4. Banerjee, S., Olvera-Cravioto, M.: PageRank asymptotics on directed preferential attachment networks. Ann. Appl. Probab. **32**(4), 3060–3084 (2022). https://doi.org/10.1214/21-aap1757
5. Barabási, A.L., Albert, R.: Emergence of scaling in random networks. Science **286**(5439), 509–512 (1999). https://doi.org/10.1126/science.286.5439.509

6. Brin, S., Page, L.: The anatomy of a large-scale hypertextual web search engine. Comput. Netw. ISDN Syst. **30**(1), 107–117 (1998). https://doi.org/10.1016/S0169-7552(98)00110-X. proceedings of the Seventh International World Wide Web Conference
7. Chen, N., Litvak, N., Olvera-Cravioto, M.: Generalized PageRank on directed configuration networks. Random Struct. Algorithms **51**(2), 237–274 (2017). https://doi.org/10.1002/rsa.20700
8. Henning, F., van der Hofstand, R., Litvak, N.: Power-law hypothesis for PageRank on undirected graphs. Preprint (2024)
9. van der Hofstad, R.: Random Graphs and Complex Networks, vol. 2. Cambridge Series in Statistical and Probabilistic Mathematics, Cambridge University Press (2024). https://doi.org/10.1017/9781316795552
10. Lawrence, S., Giles, C.L.: Searching the world wide web. Science **280**(5360), 98–100 (1998). http://www.jstor.org/stable/2895232
11. Lee, J., Olvera-Cravioto, M.: Pagerank on inhomogeneous random digraphs. Stochast. Processes Appl. **130**(4), 2312–2348 (2020). https://doi.org/10.1016/j.spa.2019.07.002
12. Leskovec, J., Kleinberg, J., Faloutsos, C.: Graphs over time: densification laws, shrinking diameters and possible explanations. In: Proceedings of the Eleventh ACM SIGKDD International Conference on Knowledge Discovery in Data Mining, pp. 177–187 (2005). https://www.cs.cmu.edu/~jure/pubs/powergrowth-kdd05.pdf
13. Leskovec, J., Mcauley, J.: Learning to discover social circles in ego networks. In: Advances in Neural Information Processing Systems, vol. 25 (2012). http://i.stanford.edu/~julian/pdfs/nips2012.pdf
14. Olvera-Cravioto, M.: PageRank's behavior under degree correlations. Ann. Appl. Probab. **31**(3), 1403–1442 (2021). https://doi.org/10.1214/20-aap1623
15. Pandey, M.: Centrality Measures and Connectivity Properties in Large Networks: Who is the most influential in a network? Ph.d. thesis 1 (research tu/e / graduation tu/e), Mathematics and Computer Science (2025). https://pure.tue.nl/ws/portalfiles/portal/345650626/20250108_Pandey_hf.pdf, proefschrift
16. Rozemberczki, B., Allen, C., Sarkar, R.: Multi-scale attributed node embedding. J. Complex Netw. **9**(2), cnab014 (2021)

Degrees in Preferential Attachment Networks with an Anomaly

Qiu Liang[✉], Remco van der Hofstad, and Nelly Litvak

Department of Mathematics and Computer Science, Eindhoven University
of Technology, Eindhoven, The Netherlands
{q.liang2,r.w.v.d.hofstad,n.v.litvak}@tue.nl

Abstract. We consider a preferential attachment model that incorporates an anomaly. Our goal is to understand the evolution of the network before and after the occurrence of the anomaly by studying the influence of the anomaly on the structural properties of the network. The anomaly is such that after its arrival it attracts newly added edges with fixed probability. We investigate the growth of degrees in the network, finding that the anomaly's degree increases almost linearly. We also provide a heuristic derivation for the exponent of the limiting degree distributions of ordinary vertices, and study the degree growth of the oldest vertex. We show that when the anomaly enters early, the degree distribution is altered significantly, while a late anomaly has minimal impact. Our analysis provides deeper insights into the evolution of preferential attachment networks with an anomalous vertex.

Keywords: Preferential attachment network · Anomaly · Degree Structure · Dynamic network

1 Introduction

Dynamic network models, where nodes and edges appear or disappear over time, attracted a lot of attention in the network science literature. Among these models, the Preferential Attachment (PA) network, as presented by Barabási and Albert [2], has been particularly influential, because it explains the emergence of scale-free property in networks through the 'rich-get-richer' phenomenon. Specifically, the probability that a new vertex connects to an existing vertex is proportional to the degree of the existing vertex.

One extension of the standard PA network is the *superstar model* [7], which is used to analyze key features of retweet networks. In this model, the initial vertex in the network is a *superstar*. At each time step, a newly added vertex connects to the superstar with probability p, or to one of the non-superstar vertices with probability $1 - p$ according to the preferential attachment rule. The superstar alters the dynamics and the resulting degree distribution. It was shown in [7] that the non-superstar vertices follow a power-law degree distribution, though with a modified exponent $3 + \frac{p}{1-p}$. Additionally, the maximal degree of the non-superstar vertices is likewise affected; it grows slower with the network size. In this paper, we consider a PA model with an anomaly. Our anomaly is similar to the *superstar*, but it may arrive at any point of time.

Since the anomaly alters the network dynamics, our work is closely related to the line of research on PA models with a *change point*, where a parameter of the PA model changes at some point of time. For instance, [1,6] investigate methods to identify the change point in preferential attachment trees via embedding the discrete time tree in a continuous-time branching process and studying the proportion of leaves, while [5] proposes an approach based on the fraction of vertices with minimal degree to detect a late change point. Furthermore, [8] applies the likelihood ratio technique to estimate the change point in a PA model, and extend the method to detect multiple change points via screening and ranking, as well as binary segmentation.

While *change point* detection addresses abrupt changes in the parameters of the network dynamics, here we instead focus on how structural properties evolve when a single anomalous vertex enters a PA network. To explore this, we suggest a new model incorporating an anomaly. Our model is an extension of the *superstar model*, and the two models are highly similar when the anomaly coincides with the *initial* vertex. Our main contributions are as follows:

▷ We propose a PA network with an anomaly. The network evolves according to the standard preferential attachment rule until the anomaly enters. Once the anomaly appears in the network, it attracts newly added edges with a fixed probability, plus a probability that depends on its current degree as in the normal PA dynamics.

▷ We compute the mean degree of the anomaly as a function of network size, and study the mean degree of other vertices and their convergence.

▷ We provide a heuristic derivation of the limiting degree distribution of the ordinary vertices when the anomaly arrives in various different stages of the network's evolution.

Our results serve as a first step towards the understanding of anomaly detection in PA networks. In the final section, we provide an outlook on this detection problem.

2 Model Description

We start by introducing the preferential attachment network without an anomaly. The model constructs a graph sequence $(G_t)_{t \geq 2}$ such that each graph G_t is formed by adding one new vertex and m edges connecting the new vertex to existing vertices, where $m \geq 1$. Let $G_t = (V_t, E_t)$, where $V_t = \{v_1, \cdots, v_t\}$ and $E_t \subseteq \{\{v, w\} : v, w \in V_t\}$. The *initial* graph G_1 consists of a single vertex v_1 and m self-loops. For $t > 1$, no self-loops are present. We treat the process of connecting each edge from a newly introduced vertex to an existing vertex in the network as an individual step. Specifically, let $G_{t,j}$ denote the network after the jth edge of a newly added vertex v_t connects to an existing vertex

$v_i \in \{v_1, v_2, \cdots, v_{t-1}\}$, where $j \in [m] \equiv \{1, \ldots, m\}$. We let $D_i(t, j)$ denote the degree of vertex i in $G_{t,j}$, and introduce a *fitness* parameter $\delta > -m$. Further, we define $G_t = G_{t,m} = G_{t+1,0}$, and $D_i(t) = D_i(t, m)$. For $t > 1$, $j \in [m]$, we define the attachment rule for the jth edge linking to the vertex $v_i \in \{v_1, v_2, \cdots, v_{t-1}\}$ as defined in [3,4], and studied further in [5]:

$$P(v_{t,j} \to v_i \mid G_{t,j-1}) = \frac{D_i(t, j-1) + \delta}{2m(t-1) + (t-1)\delta + (j-1)}. \qquad (1)$$

Assume that an anomaly occurs at some time τ satisfying $1 < \tau < t$, where the attachment rule changes after the anomaly has occurred. We denote the anomaly by the vertex v_τ. After τ, each new edge connects to the anomaly with probability

$$p \approx \frac{\beta}{2m + \beta + \delta}, \qquad (2)$$

where $\beta > 0$ is a parameter of the model. Otherwise, with complementary probability, the edge connects to any existing vertex, including the anomaly, by following the usual PA rule. Formally, the dynamics of the model *at step t* are as follows:

(I) If $t < \tau$, the anomalous vertex has not occurred in the graph G_t, the attachment rule of G_t is the same as (1).
(II) If $1 < \tau \leq t$, the evolution rule changes as follows:

$$P(v_{t,j} \to v_i \mid G_{t,j-1}) = \begin{cases} \dfrac{D_i(t, j-1) + \delta}{(t-1)(2m + \beta + \delta) + j - 1} & \text{if } i \neq \tau, \\ \dfrac{(t-1)\beta + D_\tau(t, j-1) + \delta}{(t-1)(2m + \beta + \delta) + j - 1} & \text{if } i = \tau. \end{cases} \qquad (3)$$

To explain the rationale behind (3), suppose $m = 1$, hence $j = 1$ is the only edge of vertex v_t. Then

$$\frac{(t-1)\beta + D_\tau(t-1) + \delta}{(t-1)(2 + \beta + \delta)} = p + (1-p) \frac{D_\tau(t-1) + \delta}{(t-1)(2 + \delta)},$$

with $p = \frac{\beta}{2 + \beta + \delta}$. Hence, we can think of our connection rule as connecting with probability p to the anomaly, and proportional to the degrees (including the anomaly) with probability $1 - p$. This explains the choice in (2), which is *exactly* correct for $m = 1$. To make (2) *exactly* correct for $m > 1$, we have to choose $p = p_{\beta,t,j}$ slightly differently, so that

$$p_{\beta,t,j} + (1 - p_{\beta,t,j}) \frac{D_\tau(t, j-1) + \delta}{(t-1)(2m + \delta) + j - 1}$$
$$= \frac{p_{\beta,t,j}[(t-1)(2m + \delta) + (j-1)] + (1 - p_{\beta,t,j})(D_\tau(t, j-1) + \delta)}{(t-1)(2m + \delta) + j - 1}$$
$$= \frac{(t-1)\beta + D_\tau(t, j-1) + \delta}{(t-1)(2m + \beta + \delta) + j - 1},$$

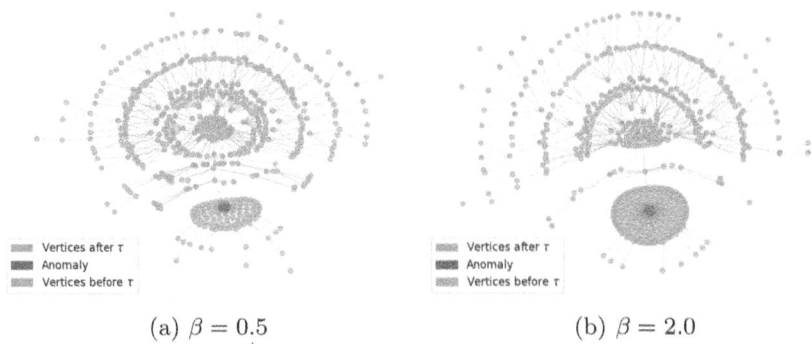

Fig. 1. Examples of PA networks with an anomaly. Here $t = 500, \tau = 200, \delta = 0, m = 1$.

which leads to $p_{\beta,t,j} \approx p$ in (3). Since the precise form of (3) is a little simpler, we choose to work with this parameterization instead to simplify the formulas.

Figure 1 illustrates an example of a PA network with an anomaly. We see that a large number of edges connect to the anomaly, yet a positive proportion of the edges of vertices arriving after τ are attached to ordinary vertices. Our goal is to study the structural properties of a PA network with an anomaly and the asymptotic degree distribution for different types of vertices.

3 The Growth of the Degrees in G_t

In this section, we analyze the growth of the degrees in G_t, both for the anomaly (Sect. 3.1) and the ordinary vertices (Sects. 3.2, 3.3). We follow the approach in [9, Chapter 8], adapted to our model with an anomaly.

3.1 Expected Degree of the Anomaly

We start by investigating the expected degree of the anomaly:

Proposition 1 (Degree of the anomaly). *Consider an anomaly that occurs at time τ, where $1 < \tau < t$, and follows the attachment rule (3) with given $\delta > -m$, and $\beta > 0$. Then*

$$\mathbb{E}[D_\tau(t) + \delta] = \frac{m\beta t}{m + \beta + \delta} + c_0 \frac{\Gamma(t + \frac{m}{2m+\beta+\delta})\Gamma(\tau)}{\Gamma(t)\Gamma(\tau + \frac{m}{2m+\beta+\delta})}. \quad (4)$$

where $c_0 = m + \delta - \frac{m\beta\tau}{m+\beta+\delta}$. In particular,

$$\lim_{\beta \to \infty} \mathbb{E}[D_\tau(t) + \delta] = (t - \tau + 1)m + \delta. \quad (5)$$

Proof. Recall that $\mathbb{E}[D_\tau(\tau) + \delta] = m + \delta$, for $1 < \tau < t$. Based on (3), the expected degree of the anomaly satisfies the recursion

$$\mathbb{E}[D_\tau(t,m) + \delta \mid D_\tau(t, m-1)]$$
$$= D_\tau(t, m-1) + \delta + \mathbb{E}[D_\tau(t,m) - D_\tau(t, m-1) \mid G_{t,m-1}]$$
$$= D_\tau(t, m-1) + \delta + \frac{(t-1)\beta + D_\tau(t, m-1) + \delta}{(t-1)(2m+\beta+\delta) + (m-1)}$$
$$= (D_\tau(t, m-1) + \delta)\left(1 + \frac{1}{(t-1)(2m+\beta+\delta) + (m-1)}\right)$$
$$+ \frac{(t-1)\beta}{(t-1)(2m+\beta+\delta) + (m-1)}.$$

Taking expectation, and solving the recursion, gives that

$$\mathbb{E}[D_\tau(t,m) + \delta] = (m+\delta) \prod_{k_2=\tau}^{t-1} \prod_{k_1=0}^{m-1} \left(1 + \frac{1}{k_2(2m+\beta+\delta) + k_1}\right) + C^m(\beta, \delta, t)$$

$$= (m+\delta) \prod_{k_2=\tau}^{t-1} \frac{k_2 + \frac{m}{2m+\beta+\delta}}{k_2} + C^m(\beta, \delta, t)$$

$$= (m+\delta) \frac{\Gamma(t + \frac{m}{2m+\beta+\delta})\Gamma(\tau)}{\Gamma(\tau + \frac{m}{2m+\beta+\delta})\Gamma(t)} + C^m(\beta, \delta, t), \qquad (6)$$

where $C^m(\beta, \delta, t)$ is a function of β, t and δ, that, with $c = 2m + \beta + \delta$, equals

$$C^m(\beta, \delta, t) = \sum_{k=0}^{m-1} \left[\frac{(t-1)\beta}{(t-1)c+k} \times \frac{(t-1)c+m}{(t-1)c+k+1}\right]$$
$$+ \frac{(t-1)c+m}{(t-1)c} \times \sum_{k=0}^{m-1}\left[\frac{(t-2)\beta}{(t-2)c+k} \times \frac{(t-2)c+m}{(t-2)c+k+1}\right] + \cdots$$
$$+ \prod_{t_1=\tau+1}^{t-1} \frac{t_1 c + m}{t_1 c} \times \sum_{k=0}^{m-1}\left[\frac{\tau\beta}{\tau c + k} \times \frac{\tau c + m}{\tau c + k + 1}\right].$$

For each time step, by the telescoping sum identity,

$$\sum_{k=0}^{m-1} \frac{t_1 \beta (t_1 c + m)}{(t_1 c + k)(t_1 c + k + 1)} = t_1 \beta (t_1 c + m) \sum_{k=0}^{m-1}\left[\frac{1}{(t_1 c + k)} - \frac{1}{(t_1 c + k + 1)}\right]$$
$$= t_1 \beta (t_1 c + m)\left(\frac{1}{t_1 c} - \frac{1}{t_1 c + m}\right) = \frac{m\beta}{c}.$$

Thus,

$$C^m(\beta,\delta,t) = \frac{m\beta}{c}\left[1 + \frac{(t-1)c+m}{(t-1)c} + \cdots + \prod_{t_1=\tau+1}^{t-1}\frac{t_1c+m}{t_1c}\right]$$

$$= \frac{m\beta}{c}\sum_{t_1=\tau+1}^{t}\frac{\Gamma(t+\frac{m}{c})\Gamma(t_1)}{\Gamma(t_1+\frac{m}{c})\Gamma(t)}.$$

Using properties of the gamma function, we can rewrite

$$C^m(\beta,\delta,t) = \frac{m\beta\Gamma(t+\frac{m}{c})}{c\Gamma(t)}\left(\frac{1}{\frac{m}{c}-1}\right) \times \sum_{t_1=\tau+1}^{t}\left(\frac{\Gamma(t_1)}{\Gamma(t_1-1+\frac{m}{c})} - \frac{\Gamma(t_1+1)}{\Gamma(t_1+\frac{m}{c})}\right)$$

$$= \frac{m\beta}{m+\beta+\delta}\frac{\Gamma(t+\frac{m}{2m+\beta+\delta})}{\Gamma(t)}\left(\frac{\Gamma(t+1)}{\Gamma(t+\frac{m}{2m+\beta+\delta})} - \frac{\Gamma(\tau+1)}{\Gamma(\tau+\frac{m}{2m+\beta+\delta})}\right)$$

$$= \frac{m\beta t}{m+\beta+\delta} - \frac{m\beta\tau}{m+\beta+\delta}\frac{\Gamma(t+\frac{m}{2m+\beta+\delta})\Gamma(\tau)}{\Gamma(t)\Gamma(\tau+\frac{m}{2m+\beta+\delta})}. \tag{7}$$

Substituting (7) into (6), we get (4). In particular, $\frac{m\beta}{m+\beta+\delta} \to m$ and $\frac{m}{2m+\beta+\delta} \to 0$ as $\beta \to \infty$, so that

$$\lim_{\beta\to\infty}\mathbb{E}[D_\tau(t)+\delta] = mt + (m+\delta-m\tau),$$

which implies that all the incoming edges are expected to connect to the anomaly, and we get (5).

For general $j \in [m]$, we can extend (4) to find $\mathbb{E}[D_\tau(t,j)+\delta]$. Applying the recursive approach,

$$\mathbb{E}[D_\tau(t,j)+\delta \mid D_\tau(t,j-1)]$$
$$= D_\tau(t,j-1) + \delta + \frac{(t-1)\beta + D_\tau(t,j-1) + \delta}{(t-1)(2m+\beta+\delta)+j-1}$$
$$= (D_\tau(t-1,m)+\delta) \times \frac{t-1+\frac{j}{2m+\beta+\delta}}{t-1} + \frac{\beta j}{2m+\beta+\delta},$$

and taking expectation, we have

$$\mathbb{E}[D_\tau(t,j)+\delta] = \mathbb{E}[D_\tau(t-1,m)+\delta]\left(1 + \frac{j}{(t-1)(2m+\beta+\delta)}\right) + \frac{\beta j}{2m+\beta+\delta}.$$

We obtain (4) by solving this recursion. □

Figure 2 shows the growth of $D_\tau(t)$ over time, with our theoretical result for $\mathbb{E}[D_\tau(t)+\delta]$ depicted as yellow line. It is very interesting that the coefficient of the linear growth, $\frac{\beta}{m+\beta+\delta}$ is larger than the probability $\frac{\beta}{2m+\beta+\delta}$ in (3) that an edge attaches itself to an anomaly. Indeed, our assumption that vertices may attach to the anomaly also through the PA mechanism, has increased the rate of growth

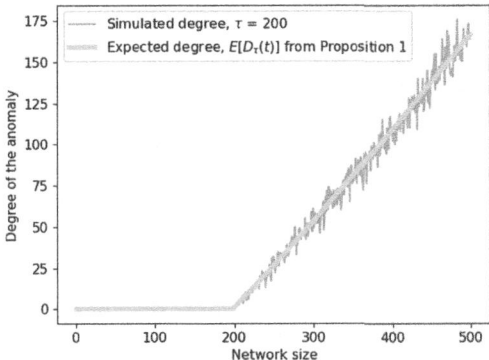

Fig. 2. The degree of the anomaly as a function of time t. The parameters used are $\tau = 200, \delta = 0, \beta = 2.0, m = 1$.

(rather than, say, giving rise to an extra polynomial term as we conjectured at the beginning). To further explore the relation between $D_\tau(t)$ and the network size, we assume that $t = a\tau$, $a \geq 1$. By Stirling's formula,

$$\frac{\Gamma(t+a)}{\Gamma(t)} = t^a(1 + O(1/t)), \qquad (8)$$

when $t \to \infty$ and a is fixed (see, e.g., [9, (8.3.9)]). This approximation allows us to simplify the formula of expected degree of v_τ. Indeed, applying it to (4) gives

$$f(a) = \lim_{\tau \to \infty} \frac{\mathbb{E}[D_\tau(a\tau) + \delta]}{a\tau} = \frac{m\beta}{m + \beta + \delta}\left(1 - a^{-\frac{m+\beta+\delta}{2m+\beta+\delta}}\right),$$

and

$$f(1) = 0, \qquad \lim_{a \to \infty} f(a) = \frac{m\beta}{m + \beta + \delta},$$

$$f'(a) = \frac{m\beta}{2m + \beta + \delta} a^{-\frac{m+\beta+\delta}{2m+\beta+\delta} - 1}, \qquad f'(1) = \frac{m\beta}{2m + \beta + \delta}.$$

We see that larger values of a, corresponding to an earlier anomaly, result in a larger deviation from the linear growth of $\mathbb{E}[D_\tau(a\tau)]$.

3.2 Expected Degree of the Ordinary Vertices

Despite the presence of an anomaly, ordinary vertices continue to receive edges based on the preferential attachment mechanism. We apply the recursive approach from [9, Section 8.3] to calculate their expected degrees, as the anomaly does not alter the edge assignment mechanism for these vertices. Regarding the attachment function, it differs for $i \leq t \leq \tau$ and $t > \tau$. Therefore, we will analyze the following two scenarios:

(1) If $i < \tau$, for $t < \tau$, the anomaly has not yet occurred, the expected degree of v_i is

$$\mathbb{E}[D_i(G_t) + \delta] = \begin{cases} (2m+\delta)\dfrac{\Gamma(t + \frac{m}{2m+\delta})}{\Gamma(1 + \frac{m}{2m+\delta})\Gamma(t)}, & i = 1, \\ (m+\delta)\dfrac{\Gamma(t + \frac{m}{2m+\delta})\Gamma(i)}{\Gamma(i + \frac{m}{2m+\delta})\Gamma(t)}, & 1 < i < \tau. \end{cases} \qquad (9)$$

See, e.g., [9, Exercise 8.14] for the case where $m = 1$. Remarkably, for the model chosen here, the formula for $m > 1$ is actually quite nice.

For $t > \tau$, the attachment rule is changed after the anomaly appears. Then, the expected degree of v_i changes into

$$\mathbb{E}[D_i(t) + \delta] = \begin{cases} (2m+\delta)\dfrac{\Gamma(\tau + \frac{m}{2m+\delta})\Gamma(t + \frac{m}{2m+\beta+\delta})}{\Gamma(1 + \frac{m}{2m+\delta})\Gamma(\tau + \frac{m}{2m+\beta+\delta})\Gamma(t)}, & i = 1, \\ (m+\delta)\dfrac{\Gamma(\tau + \frac{m}{2m+\delta})\Gamma(t + \frac{m}{2m+\beta+\delta})\Gamma(i)}{\Gamma(i + \frac{m}{2m+\delta})\Gamma(\tau + \frac{m}{2m+\beta+\delta})\Gamma(t)}, & 1 < i < \tau. \end{cases}$$
(10)

(2) If $i > \tau$, for $t > \tau$, the expected degree of v_i is

$$\mathbb{E}[D_i(t) + \delta] = (m+\delta)\frac{\Gamma(t + \frac{m}{2m+\beta+\delta})\Gamma(i)}{\Gamma(i + \frac{m}{2m+\beta+\delta})\Gamma(t)}. \qquad (11)$$

3.3 Convergence of Degrees for Ordinary Vertices

In [9, Section 8.3], it is proved that the degree of vertices in a standard PA network scales as $t^{\frac{1}{2+\delta}}$ when $m = 1$, and as $t^{\frac{1}{2+\delta/m}}$ when $m \geq 2$ (see [9, Exercise 8.13]). In our model, if $t < \tau$, the anomaly is not included in G_t, so that again the degrees of vertices are of the order $t^{\frac{1}{2+\delta/m}}$. When $t > \tau$, it is necessary to analyze the convergence of the degrees of vertices separately.

For the vertices added before τ, we consider the sequence $(M_i^{(1)}(t))_{t \geq i}$ given by

$$M_i^{(1)}(t) = \frac{D_i(t) + \delta}{\mathbb{E}[D_i(t) + \delta]}.$$

It is easy to see that $(M_i^{(1)}(t))_{t \geq i}$ is a non-negative martingale. Indeed, since $m \leq D_i(t) < 2mt$, we have $\mathbb{E}[|M_i^{(1)}(t)|] < \infty$. Computing the conditional expectation, we get

$$\mathbb{E}[M_i^{(1)}(t+1) \mid M_i^{(1)}(t)] = \mathbb{E}[M_i^{(1)}(t+1) \mid D_i(t)]$$
$$= \frac{\mathbb{E}[D_i(t+1) + \delta \mid D_i(t)]}{\mathbb{E}[D_i(t+1) + \delta]}$$
$$= \frac{(D_i(t) + \delta)}{\mathbb{E}[D_i(t+1) + \delta]} \prod_{j=1}^{m} \left(1 + \frac{1}{t(2m + \beta + \delta) + j - 1}\right)$$
$$= \frac{D_i(t) + \delta}{\mathbb{E}[D_i(t+1) + \delta]} \cdot \frac{t + \frac{m}{2m+\beta+\delta}}{t}$$
$$= \frac{D_i(t) + \delta}{\mathbb{E}[D_i(t) + \delta]} = M_i^{(1)}(t),$$

since also

$$\mathbb{E}[D_i(t+1) + \delta] = \mathbb{E}[D_i(t) + \delta] \frac{t + \frac{m}{2m+\beta+\delta}}{t}. \tag{12}$$

Thus, $(M_i^{(1)}(t))_{t \geq i}$ is a non-negative martingale with respect to $(G_t)_{t \geq i}$. According to the martingale convergence theorem, $M_i^{(1)}(t)$ converges almost surely to a limiting random variable as $t \to \infty$ [9, Theorem 2.24], consequently, the result can be extended to establish the convergence of degrees when $i < \tau$. By the Stirling's formula, for sufficiently large t and τ,

$$\frac{D_i(t) + \delta}{\left(\frac{t}{\tau}\right)^{\frac{m}{2m+\beta+\delta}} \tau^{\frac{m}{2m+\delta}}} = M_i^{(1)}(t) \frac{(d+\delta)\Gamma(i)}{\Gamma(i + \frac{m}{2m+\delta})}(1 + o(1))$$
$$\xrightarrow{a.s.} \frac{(d+\delta)\Gamma(i)}{\Gamma(i + \frac{m}{2m+\delta})} \xi_i^{(1)},$$

where $\xi_i^{(1)}$ is the almost sure limit of $M_i^{(1)}(t)$, and $d = 2m$ for $i = 1$, and $d = m$ for $i > 1$. Thus, when $i < \tau$, $\frac{D_i(t)+\delta}{\left(\frac{t}{\tau}\right)^{\frac{m}{2m+\beta+\delta}} \tau^{\frac{m}{2m+\delta}}}$ converges almost surely as $t \to \infty$ and $\tau \to \infty$.

Similarly, for the vertices added after τ, let $(M_i^{(2)}(t))_{t \geq i}$ be given by

$$M_i^{(2)}(t) = \frac{D_i(t) + \delta}{\mathbb{E}[D_i(t) + \delta]}, \qquad i > \tau.$$

Following the previously described steps again, we get $(M_i^{(2)}(t))_{t \geq i}$ is a non-negative martingale, and when t is large enough,

$$\frac{D_i(t) + \delta}{t^{\frac{m}{2m+\beta+\delta}}} = M_i^{(2)}(t) \frac{(m+\delta)\Gamma(i)}{\Gamma(i + \frac{m}{2m+\beta+\delta})}(1 + o(1))$$
$$\xrightarrow{a.s.} \frac{(m+\delta)\Gamma(i)}{\Gamma(i + \frac{m}{2m+\beta+\delta})} \xi_i^{(2)},$$

where $\xi_i^{(2)}$ is the almost sure limit of $M_i^{(2)}(t)$. Thus, when $i > \tau$, $\frac{D_i(t)+\delta}{t^{\frac{m}{2m+\beta+\delta}}}$ converges almost surely.

4 Heuristic Derivation of the Limiting Degree Distribution for Ordinary Vertices

In this section, we aim to investigate the limiting degree distribution of the ordinary vertices in a PA network containing an anomaly. For the standard PA network, the exponent of power law degree is $3 + \frac{\delta}{m}$. The rigorous proof in [9, Chapter 8] strongly relies on the recursive relation between the fractions of vertices with degree k. However, it is hard to adapt this approach, due to the alteration in the recursion after the appearance of an anomaly. Consequently, the rigorous derivation remains an open problem for future research. Here, we present a heuristic argument [11] that yields results consistent with our numerical simulations.

The main idea is that if we can find the range of vertex index i such that the average degree of vertex v_i falls in $(k-0.5, k+0.5)$, then the limiting fraction p_k of vertices of degree k can be evaluated as that range divided by t, as $t \to \infty$. We will explore the power-law degree distribution in different scaling regimes for the arrival time of anomaly τ as a function of time t. Specifically, we consider three cases: when the anomaly arrives late, mid-way or early.

4.1 Late Anomaly: $\tau = t - t^\gamma, \gamma \in (0,1)$

If the arrival time of v_τ is quite late, for example when $\tau = t - t^\gamma$, where $\gamma \in (0,1)$, then only a vanishing fraction of vertices arrives after the anomaly. Therefore, the degree distribution will be defined by the vertices that arrived before the anomaly. When $i < \tau$, the expected degree of v_i is (10), here we only consider $i > 1$, as t and τ are large enough, by the Stirling formula, $\mathbb{E}[D_i(t) + \delta]$ can be approximated as

$$\mathbb{E}[D_i(t)+\delta] = (m+\delta)\left(\frac{t}{t-t^\gamma}\right)^{\frac{m}{2m+\beta+\delta}}\left(\frac{t-t^\gamma}{i}\right)^{\frac{m}{2m+\delta}}(1+o(1)), \quad i < \tau. \quad (13)$$

Assume that $\mathbb{E}[D_i(t) + \delta]$ falls in the interval $(k + \delta - 0.5, k + \delta + 0.5)$, then i should be in

$$\left((m+\delta)^{2+\frac{\delta}{m}}\frac{(1-t^{\gamma-1})^{\frac{\beta}{2m+\beta+\delta}}t}{(k+\delta+0.5)^{2+\frac{\delta}{m}}}, (m+\delta)^{2+\frac{\delta}{m}}\frac{(1-t^{\gamma-1})^{\frac{\beta}{2m+\beta+\delta}}t}{(k+\delta-0.5)^{2+\frac{\delta}{m}}}\right).$$

Let L_1 be the length of interval,

$$L_1 = (m+\delta)^{2+\frac{\delta}{m}}(1-t^{\gamma-1})^{\frac{\beta}{2m+\beta+\delta}}t\left(\frac{1}{(k+\delta-0.5)^{2+\frac{\delta}{m}}} - \frac{1}{(k+\delta+0.5)^{2+\frac{\delta}{m}}}\right)$$

$$= (m+\delta)^{2+\frac{\delta}{m}}(1-t^{\gamma-1})^{\frac{\beta}{2m+\beta+\delta}}t\cdot\frac{k^{2+\frac{\delta}{m}}\left[\left(1+\frac{\delta+0.5}{k}\right)^{2+\frac{\delta}{m}} - \left(1-\frac{0.5-\delta}{k}\right)^{2+\frac{\delta}{m}}\right]}{k^{4+\frac{2\delta}{m}}\left(1+\frac{\delta+0.5}{k}\right)^{2+\frac{\delta}{m}}\left(1-\frac{0.5-\delta}{k}\right)^{2+\frac{\delta}{m}}}.$$

Next we apply the binomial theorem to simplify the expression of L_1, to obtain, as $k \to \infty$,

$$L_1 = \left(2 + \frac{\delta}{m}\right)(m+\delta)^{2+\frac{\delta}{m}}(1-t^{\gamma-1})^{\frac{\beta}{2m+\beta+\delta}}tk^{-(3+\frac{\delta}{m})}(1+o(1)),$$

so that, for $k \to \infty$,

$$p_k^{(\text{late})} = \lim_{t \to \infty} \frac{L_1}{t} = \left(2 + \frac{\delta}{m}\right)(m+\delta)^{2+\frac{\delta}{m}}k^{-(3+\frac{\delta}{m})}(1+o(1)). \qquad (14)$$

Thus, the asymptotic degree distribution remains unchanged compared to the standard PA network.

4.2 Mid-Way Anomaly: $\tau = \alpha T$, $\alpha \in (0, 1)$

Assume that the arrival time of v_τ scales linearly with t, that is, $\tau = \alpha t$, where $\alpha \in (0, 1)$. For $1 < i < \tau$, the average degree is approximated by

$$\mathbb{E}[D_i(t) + \delta] = (m+\delta)\alpha^{-\frac{m}{2m+\beta+\delta}}\left(\frac{\alpha t}{i}\right)^{\frac{m}{2m+\delta}}(1+o(1)), \qquad i < \tau. \qquad (15)$$

Similarly, for the vertex v_i with $i > \tau$,

$$\mathbb{E}[D_i(t) + \delta] = (m+\delta)\left(\frac{t}{i}\right)^{\frac{m}{2m+\beta+\delta}}(1+o(1)). \qquad (16)$$

We see that the average degree of the vertices born after τ grows slower than that of the vertices that arrived before τ. Moreover, the right-hand side of (16) is bounded, and thus it cannot fall into the interval $(k + \delta - 0.5, k + \delta + 0.5)$ for large k. Therefore, we perform derivations for the vertices that have arrived before time τ as the higher order probability to achieve a large degree k. Repeat the steps in Sect. 4.1, we get, as $k \to \infty$,

$$p_k^{(\text{mid-way})} = \left(2 + \frac{\delta}{m}\right)(m+\delta)^{2+\frac{\delta}{m}}\alpha^{\frac{\beta}{2m+\beta+\delta}}k^{-(3+\frac{\delta}{m})}(1+o(1)). \qquad (17)$$

From this formula we see that the vertices arriving before τ follow the power law distribution with exponent $3 + \frac{\delta}{m}$. This is the same power-law exponent as in the standard PA network. However, we get a factor $\alpha^{\frac{\beta}{2m+\beta+\delta}}$ in front. This conforms with the intuition that the anomaly slows down the degree growth of high-degree vertices. Moreover, this factor decreases with β, as larger β increases the effect of the anomaly.

4.3 Early Anomaly: $\tau = t^\gamma$, $\gamma \in (0, 1)$

Suppose that the anomaly arrives quite early at $\tau = t^\gamma$, where $\gamma \in (0, 1)$. Then the fraction of vertices born before τ among all vertices is vanishing. Therefore, we investigate the behavior of vertices born after τ. Repeating the steps in

Sect. 4.1, we obtain an asymptotic degree distribution given by

$$p_k^{(\text{early})} = \left(2 + \frac{\beta}{m} + \frac{\delta}{m}\right)(m+\delta)^{2+\frac{\beta}{m}+\frac{\delta}{m}} k^{-(3+\frac{\beta}{m}+\frac{\delta}{m})}(1+o(1)), \ k \to \infty. \quad (18)$$

The power-law exponent of the degree distribution changes to $3 + \frac{\beta}{m} + \frac{\delta}{m}$. We see that the anomaly has altered the power-law exponent, as also observed in the superstar model [7].

4.4 Comparison to the Empirical Degree Distribution

Figure 3, 4 and 5 show the empirical and the theoretically predicted degree distributions for the PA network with the late anomaly, the mid-way anomaly, and the early anomaly, respectively.

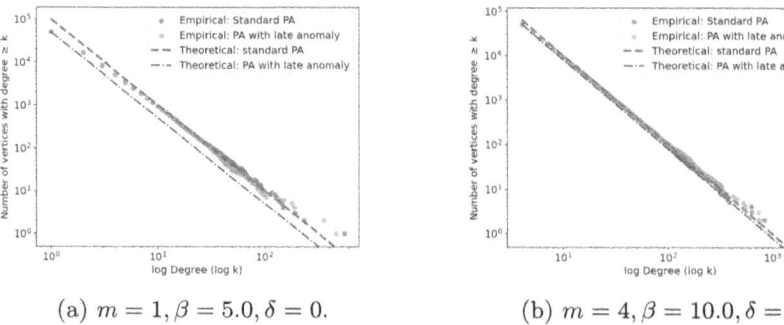

(a) $m = 1, \beta = 5.0, \delta = 0$. (b) $m = 4, \beta = 10.0, \delta = 0$.

Fig. 3. The complementary cumulative degree distribution for PA network with late anomaly, parameters are $t = 50000, \tau = 49950, \gamma = 0.3615$.

Generally, we see that our computations correctly predict the slope, but the multiplicative factor may deviate from the experiments. In the future, a rigorous derivation for the mid-way and early anomaly is needed. The late anomaly is equivalent to standard PA network but we plot the line that we derived in (14) to show the difference between the correct multiplicative factor and that resulting from the heuristic derivation.

In Fig. 3 we see that when the anomaly arrives near the end of the network's growth, the proportion of vertices with degree at least k is close to that of the standard PA network. The straight line derived from (14) (dashed blue line) has the correct slope, but is slightly different from the empirical degree distribution. The dashed red line, based on the more precise formula (8.4.11) in [9, Chapter 8] for the standard preferential attachment model has both the slope and the multiplicative factor matching the experiments.

In Fig. 4 for the mid-way anomaly, the proportion of vertices with degree at least k is close to that for the standard PA network, but differs by a factor smaller

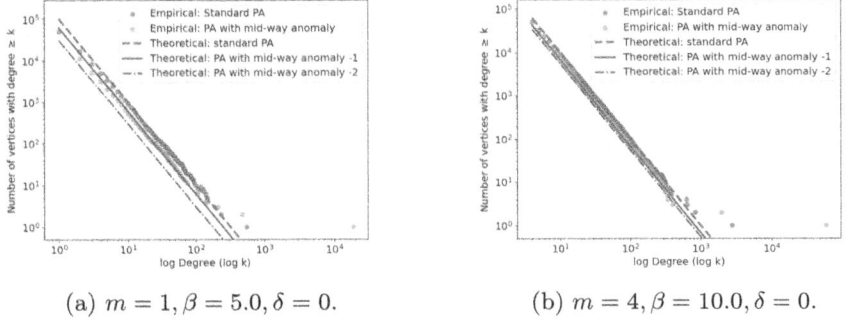

(a) $m = 1, \beta = 5.0, \delta = 0$. (b) $m = 4, \beta = 10.0, \delta = 0$.

Fig. 4. The complementary cumulative degree distribution for PA network with mid-way anomaly, parameters are $t = 50000, \tau = 25000, \alpha = 0.5$. The green line uses the formula (8.4.11) in [9, Chapter 8], multiplied by factor $\alpha^{\frac{\beta}{2m+\beta+\delta}}$ as in (17). The dashed blue line follows (17). (Color figure online)

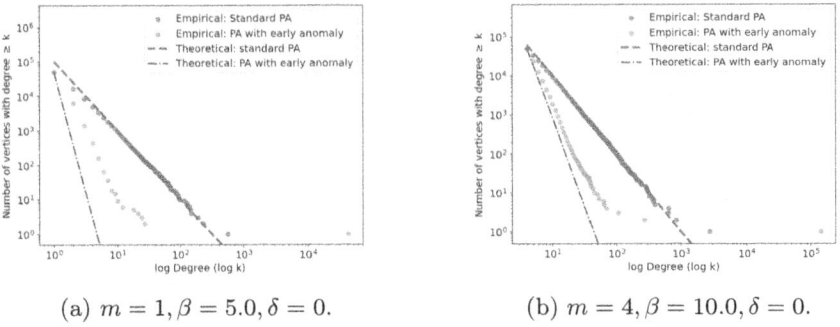

(a) $m = 1, \beta = 5.0, \delta = 0$. (b) $m = 4, \beta = 10.0, \delta = 0$.

Fig. 5. The complementary cumulative degree distribution for PA network with early anomaly, parameters are $t = 50000, \tau = 50, \gamma = 0.3615$.

than 1 as predicted by (17). Interestingly, the experiments agree with formula (8.4.11) in [9, Chapter 8] multiplied by the factor $\alpha^{\frac{\beta}{2m+\beta+\delta}}$. Thus, our heuristic derivation correctly predicts the effect of a mid-way anomaly. The visible outlier in Fig. 4 is the anomaly itself.

In Fig. 5 for the early anomaly, the proportion of vertices with degree greater than k is significantly lower, as the anomaly attracts a fraction of edges from the very beginning. However, (18) predicts an even steeper slope. We also note that the tail of the distribution deviates strongly to the right from the straight line. We will come back to this phenomenon in the next section, where we analyze the behavior of the oldest ordinary vertex.

5 Degree Growth of the Oldest Vertex

Because the existing vertices connect to a new coming vertex with a probability proportional to its degree, it is more likely that the old vertices receive more and more edges over time, and their degrees are of the order $t^{\frac{1}{2+\delta/m}}$ [9, Theorem 8.2]. We further explore how the degree of the oldest vertex behaves after the anomaly's occurrence, by computing the exponent of power-law degree of the initial vertex v_1. By Stirling's formula, as t and τ are large enough, the expected degree of the oldest vertex is

$$\mathbb{E}[D_1(t) + \delta] = (2m + \delta)\left(\frac{t}{\tau}\right)^{\frac{1}{2+\frac{\beta}{m}+\frac{\delta}{m}}} \tau^{\frac{1}{2+\frac{\delta}{m}}}(1 + o(1)).$$

In different scenarios, as $t \to \infty$, the approximate expression are shown as follows:

Late anomaly: If $\tau = t - t^\gamma, \gamma \in (0, 1)$,

$$\mathbb{E}[D_1(t) + \delta] = (2m + \delta)t^{\frac{1}{2+\frac{\delta}{m}}}(1 + o(1)).$$

Mid-way anomaly: If $\tau = \alpha t, \alpha \in (0, 1)$,

$$\mathbb{E}[D_1(t) + \delta] = (2m + \delta)\alpha^{\frac{1}{2+\frac{\delta}{m}} - \frac{1}{2+\frac{\beta}{m}+\frac{\delta}{m}}} t^{\frac{1}{2+\frac{\delta}{m}}}(1 + o(1)).$$

Early anomaly: If $\tau = t^\gamma, \gamma \in (0, 1)$,

$$\mathbb{E}[D_1(t) + \delta] = (2m + \delta)t^{\frac{\gamma}{2+\frac{\delta}{m}} + \frac{1-\gamma}{2+\frac{\beta}{m}+\frac{\delta}{m}}}(1 + o(1)).$$

Interestingly, the mid-way and the early anomaly affect the mean degree of the oldest vertex in a different way. The mid-way anomaly reduces the mean degree by a constant factor, while an early anomaly changes the exponent of t. Moreover, in the case of an early anomaly, $\mathbb{E}[D_1(t) + \delta]$ grows more quickly than $t^{\frac{1}{2+\frac{\beta}{m}+\frac{\delta}{m}}}$. The latter expression would be consistent with the power-law distribution for an early anomaly in (18) meaning that the oldest ordinary vertices have higher degrees than predicted by (18); therefore, we see many outliers to the right in Fig. 5.

6 Conclusion and Further Research

We introduced a PA network incorporating an anomaly, where the attachment rule of the anomaly is vastly different from that of the ordinary vertices. We derived insights on the growth of the degrees and their distribution in this model. Below we list some directions for further research:

(1) **Concentration of the degree of an anomalous vertex.** We see in Fig. 2 that the degree of the anomaly is close to its mean. The martingale convergence theorem is a standard way to prove such concentration. However, we could not directly apply this method to the degree of the anomaly due to the presence of a linear term in its expression.

(2) **Convergence of degree sequences.** In [9, Chapter 8], the limiting degree distribution is derived from the recursive equations for the number of vertices of degree k. However, we could not use this method because the occurrence of an anomaly changes the recursion. In previous works [1,6], the asymptotic degree distribution of PA networks with change points was rigorously derived using continuous-branching processes, which embed the growth of the PA network for $m = 1$ in continuous time. Investigating whether these techniques can be adapted to our model is a promising direction for gaining deeper insights into the interplay between anomalies and degree distributions.

(3) **Different attachment mechanism.** In this paper, we consider only the case where the anomaly attracts new edges at a *constant* additional rate. An interesting direction for future research would be to explore scenarios where the anomaly attaches new edges at rates that change over time, such as an increasing rate or for a randomly determined duration.

(4) **Anomaly detection.** Building on our results, detecting anomalies is a natural next step. Existing work [13] has utilized Lyapunov-based method to detect certain anomalous events in PA network. Our model, where the anomaly arrives at a specific point and alters the attachment mechanisms, requires a different approach. While many studies on anomaly detection in dynamic networks focus on analyzing spatial and temporal [12,14], or structural [10], features, we find it an interesting problem to detect the anomaly based only on the history of G_t. Our initial attempts show that this problem is more challenging than one could expect given how strong our anomaly is. We hope to report on the progress in the near future.

Acknowledgments. The work of QL was supported by the China Scholarship Council. The work of RvdH and NL is supported by the Netherlands Organisation for Scientific Research (NWO) through the Gravitation NETWORKS grant 024.002.003, and by the National Science Foudation under Grant No. DMS-1928930 while the authors were in residence at the Simons Laufer Mathematical Sciences Institute in Berkeley, California, during the Spring semester 2025.

References

1. Banerjee, S., Bhamidi, S., Carmichael, I.: Fluctuation bounds for continuous time branching processes and evolution of growing trees with a change point. Ann. Appl. Probab. **33**(4), 2919–2980 (2023)
2. Barabási, A.L., Albert, R.: Emergence of scaling in random networks. Science **286**(5439), 509–512 (1999)
3. Berger, N., Borgs, C., Chayes, J., Saberi, A.: Asymptotic behavior and distributional limits of preferential attachment graphs. Ann. Probab. **42**(1), 1–40 (2014)
4. Berger, N., Borgs, C., Chayes, J., Saberi, A.: On the spread of viruses on the internet. In: SODA 2005: Proceedings of the Sixteenth Annual ACM-SIAM Symposium on Discrete Algorithms, pp. 301–310. Society for Industrial and Applied Mathematics, Philadelphia, PA, USA (2005)

5. Bet, G., Bogerd, K., Castro, R.M., van der Hofstad, R.: Detecting a late changepoint in the preferential attachment model (2023). https://arxiv.org/abs/2310.02603
6. Bhamidi, S., Jin, J., Nobel, A.: Change point detection in network models: preferential attachment and long range dependence. Ann. Appl. Probab. **28**(1), 35–78 (2018)
7. Bhamidi, S., Steele, J.M., Zaman, T.: Twitter event networks and the Superstar model. Ann. Appl. Probab. **25**(5), 2462–2502 (2015)
8. Cirkovic, D., Wang, T., Zhang, X.: Likelihood-based inference for random networks with changepoints (2024). https://arxiv.org/abs/2206.01076
9. van der Hofstad, R.: Random Graphs and Complex Networks, vol. 1. Cambridge Series in Statistical and Probabilistic Mathematics, vol. 34. Cambridge University Press, Cambridge (2017)
10. Kipf, T.N., Welling, M.: Semi-supervised classification with graph convolutional networks. In: International Conference on Learning Representations (ICLR) (2017)
11. Litvak, N.: Randomness and structure in complex networks. Nieuw archief voor wiskunde **5**(24), 103–113 (2023)
12. Liu, Y., et al.: Anomaly detection in dynamic graphs via transformer. IEEE Trans. Knowl. Data Eng. **35**(12), 12081–12094 (2023)
13. Ruiz, D., Finke, J.: Lyapunov-based anomaly detection in preferential attachment networks. Int. J. Appl. Math. Comput. Sci. **29**(2), 363–373 (2019)
14. Zheng, L., Li, Z., Li, J., Li, Z., Gao, J.: Addgraph: anomaly detection in dynamic graph using attention-based temporal GCN. In: Proceedings of the Twenty-Eighth International Joint Conference on Artificial Intelligence, IJCAI 2019, pp. 4419–4425. International Joint Conferences on Artificial Intelligence Organization (2019)

The Artificial Benchmark for Community Detection with Outliers and Overlapping Communities (ABCD+o^2)

Jordan Barrett[1(✉)], Ryan DeWolfe[1], Bogumił Kamiński[2], Paweł Prałat[1], Aaron Smith[3], and François Théberge[4]

[1] Department of Mathematics, Toronto Metropolitan University, Toronto, Canada
{jordan.barrett,ryan.dewolfe,pralat}@torontomu.ca
[2] Decision Analysis and Support Unit, SGH Warsaw School of Economics, Warsaw, Poland
bkamins@sgh.waw.pl
[3] Department of Mathematics and Statistics, University of Ottawa, Ottawa, Canada
asmi28@uOttawa.ca
[4] Tutte Institute for Mathematics and Computing, Ottawa, Canada
theberge@ieee.org

Abstract. The Artificial Benchmark for Community Detection (**ABCD**) graph is a random graph model with community structure and power-law distribution for both degrees and community sizes. The model generates graphs similar to the well-known **LFR** model but is faster and more interpretable. In this paper, we use the underlying ingredients of the **ABCD** model, and its generalization to include outliers (**ABCD+o**), and introduce another variant for overlapping communities, **ABCD+o^2**.

Keywords: ABCD · community detection · overlapping communities · benchmark models

1 Introduction

One of the most important features of real-world networks is their community structure, as it reveals the internal organization of nodes. In social networks, communities may represent groups by interest; in citation networks, they correspond to related papers; in the Web graph, communities are formed by pages on related topics, etc. Identifying communities in a network is therefore valuable as it helps us understand the structure of the network.

Detecting communities is quite a challenging task. In fact, there is no definition of community that researchers and practitioners agree on. Still, it is widely accepted that a community should induce a graph that is denser than the global density of the network [10]. Numerous community detection algorithms have been developed over the years, using various techniques such as optimizing modularity, removing high-betweenness edges, detecting dense subgraphs, and statistical inference. We direct the interested reader to the survey [8] or one of the numerous books on network science such as [15].

Most community detection algorithms aim to find a partition of the set of nodes, that is, a collection of pairwise disjoint communities with the property that each node belongs to exactly one of them. This is a natural assumption for many scenarios. For example, most of the employees on LinkedIn work for a single employer. On the other hand, users of Instagram can belong to many social groups associated with their workplace, friends, sports, etc. Researchers might be part of many research groups. A large fraction of proteins belong to several protein complexes simultaneously. As a result, many real-world networks are better modelled as a collection of overlapping communities [22].

In the context of overlapping communities, one can distinguish two forms of overlap. In *crisp* overlap, nodes belong to communities with equal strength, whereas in *fuzzy* overlap, each node may belong to more than one community, but the strength of its membership to each community may vary. Most existing algorithms for detecting overlapping communities are crisp [11]. However, one may start with one of the crisp algorithms and then modify their outcomes to produce fuzzy overlap. Association scores, like the ones we recently proposed in [2], may be used to measure how strongly a node belongs to a community.

Unfortunately, there are very few datasets with ground-truth communities properly identified and labelled. As a result, there is a need for synthetic random graph models with community structure that resemble real-world networks to benchmark and tune clustering algorithms that are unsupervised by nature. The highly popular **LFR** (**L**ancichinetti, **F**ortunato, **R**adicchi) model [19,20] generates networks with communities and, at the same time, allows for heterogeneity in the distributions of both node degrees and of community sizes. It became a standard and extensively used method for generating artificial networks.

A similar synthetic network to **LFR**, the Artificial Benchmark for Community Detection (**ABCD**) [14] was recently introduced and implemented[1], along with a faster and multithreaded implementation[2] (**ABCDe**) [12]. Undirected variants of **LFR** and **ABCD** produce graphs with comparable properties, but **ABCD** (and especially **ABCDe**) is faster than **LFR** and can be easily tuned to allow the user to make a smooth transition between the two extremes: pure (disjoint) communities and random graphs with no community structure. Moreover, **ABCD** is easier to analyze theoretically—for example, in [13] various theoretical asymptotic properties of the are investigated, including the modularity function that, despite some known issues such as the "resolution limit" reported in [9], is an important graph property of networks in the context of community detection. In [3], some interesting and desired self-similar behaviour of the **ABCD** model is analyzed; namely, that the degree distribution of ground-truth communities is asymptotically the same as the degree distribution of the whole graph (appropriately normalized based on their sizes). Finally, the building blocks in the model are flexible and may be adjusted to satisfy different needs. Indeed, the original **ABCD** model was recently adjusted to include potential outliers (**ABCD+o**) [16] and extended to hypergraphs (**h–**

---

[1] https://github.com/bkamins/ABCDGraphGenerator.jl/.
[2] https://github.com/tolcz/ABCDeGraphGenerator.jl/.

ABCD) [17][3]. For these reasons **ABCD** is gaining recognition as a benchmark for community detection algorithms. For example, [1] used the **Adjusted Mutual Information** (**AMI**) between the partitions returned by various algorithms and the ground-truth partitions of **ABCD** and **LFR** graphs to compare 30 community detection algorithms, and mention that *while being directly comparable to LFR, ABCD offers additional benefits, including higher scalability and better control for adjusting an analogous mixing parameter.*

In this paper we extend the **ABCD+o** model further to allow for overlapping communities (**ABCD+o²**). The **LFR** model has been extended in a similar way [19], and in this model the nodes are assigned to communities based on the construction of a random bipartite graph between nodes and communities that results in (a) a small amount of overlap between almost every pair of communities, and (b) rarely any pair of communities with a large overlap. In **ABCD+o²**, we instead generate overlapping communities based on a hidden, low-dimensional geometric layer which tends to yield fewer and larger overlaps. Furthermore, the ancillary benefits of the ABCD model (an intuitive noise parameter, a fast implementation, and theoretical analysis) are still present, so this extension makes **ABCD+o²** an attractive option for benchmarking community detection algorithms.

The rest of the paper is organized as follows. In Sect. 2 we present the **ABCD+o²** model, with a full description of generating a graph in Sect. 2.5. Next, in Sect. 3 we show the properties of the model and test theoretical expectations versus simulated results. Then, in Sect. 4 we use the model to benchmark various community detection algorithms and compare their quality under different levels of noise and overlap. Finally, some concluding remarks are given in Sect. 5.

2 ABCD+o²–ABCD with Overlapping Communities and Outliers

As mentioned in the introduction, the original **ABCD** model was extended to include outliers resulting in the **ABCD+o** model. For our current needs, we extend **ABCD+o** further to allow for non-outlier nodes to belong to multiple communities, resulting in the **ABCD+o²** model, **ABCD** with **o**verlapping communities and **o**utliers.

2.1 Notation

For a given $n \in \mathbb{N} := \{1, 2, \ldots\}$, we use $[n]$ to denote the set consisting of the first n natural numbers, that is, $[n] := \{1, 2, \ldots, n\}$.

Power-law distributions will be used to generate both the degree sequence and community sizes so let us formally define it. For given parameters $\gamma \in (0, \infty)$,

---

[3] https://github.com/bkamins/ABCDHypergraphGenerator.jl.

$\delta, \Delta \in \mathbb{N}$ with $\delta \leq \Delta$, we define a truncated power-law distribution $\mathcal{P}(\gamma, \delta, \Delta)$ as follows. For $X \sim \mathcal{P}(\gamma, \delta, \Delta)$ and for $k \in \mathbb{N}$ with $\delta \leq k \leq \Delta$,

$$\mathbb{P}(X = k) = \frac{\int_k^{k+1} x^{-\gamma}\, dx}{\int_\delta^{\Delta+1} x^{-\gamma}\, dx}. \tag{1}$$

2.2 The Configuration Model

The well-known configuration model is an important ingredient of all variants of the **ABCD** models, so let us formally define it here. Suppose that our goal is to create a graph on n nodes with a given degree distribution $\mathbf{d} := (d_i, i \in [n])$, where $\mathbf{d}$ is a sequence of non-negative integers such that $m := \sum_{i \in [n]} d_i$ is even. We define a random multi-graph $\mathrm{CM}(\mathbf{d})$ with a given degree sequence known as the **configuration model** (sometimes called the **pairing model**), which was first introduced by Bollobás [5]. (See [4,24,25] for related models and results.)

We start by labelling nodes as $[n]$ and, for each $i \in [n]$, endowing node i with d_i half-edges. We then iteratively choose two unpaired half-edges uniformly at random (from the set of pairs of remaining half-edges) and pair them together to form an edge. We iterate until all half-edges have been paired. This process yields a graph $G_n \sim \mathrm{CM}(\mathbf{d})$ on n nodes, where G_n is allowed self-loops and multi-edges and thus G_n is a multi-graph.

2.3 Parameters of the ABCD+o² Model

The following ten parameters govern the **ABCD+o²** model.

Parameter	Range	Description
n	$\mathbb{N}$	Number of nodes
s_0	$\mathbb{N}$	Number of outliers
η	$[1, \infty)$	Average number of communities a non-outlier node is part of
γ	$(2, 3)$	Power-law degree distribution with exponent γ
δ	$\mathbb{N}$	Min degree as least δ
Δ	$\mathbb{N} \setminus [\delta - 1]$	Max degree at most Δ
β	$(1, 2)$	Power-law community size distribution with exponent β
s	$\mathbb{N} \setminus [\delta]$	Min community size at least s
S	$\mathbb{N} \setminus [s - 1]$	Max community size at most S
ξ	$[0, 1]$	Level of noise

Note that the ranges for γ and β can be relaxed and, more generally, any valid sequences for degrees and community sizes can be given as input to the model. However, the ranges $\gamma \in (2, 3)$ and $\beta \in (1, 2)$ were used in all previous theoretical work on the **ABCD** and **ABCD+o** models, and are suggested parameters based on the behaviour of real-world networks (see [14] for more details).

2.4 Big Picture

The **ABCD+o**^{2} model generates a random graph on n nodes with degree sequence $(d_i, i \in [n])$ and community size sequence $(s_i, i \in [L])$ following power laws with exponents γ and, respectively, β.

There are s_0 outliers and $\hat{n} = n - s_0$ non-outliers. Outliers will form their own auxiliary "community" C_0. Non-outliers will span a family of L communities $(C_j, j \in [L])$ with each non-outlier belonging to at least one of the communities. These communities will overlap (unless $\eta = 1$) so that non-outliers will belong to η communities, on average. The non-outliers, with their respective degrees, populate $(C_j, j \in [L])$ randomly with the caveat that high degree nodes cannot enter small communities.

Parameter $\xi \in [0, 1]$ dictates the amount of noise in the network. Each non-outlier node i has its degree d_i split into two parts: *community degree* y_i and *background degree* z_i ($d_i = y_i + z_i$). The goal is to get $y_i \approx (1-\xi)d_i$ and $z_i \approx \xi d_i$. However, y_i and z_i must be non-negative integers, and y_i must be split into η_i non-negative integers, one for each community. Moreover, the sum of degrees assigned to each community must be even. We achieve the first requirement by using an appropriate random rounding of $(1 - \xi)d_i/\eta_i$, and achieve the second requirement by making a few ± 1 adjustments at the end. Note that the neighbours of outliers are sampled from the entire graph, ignoring the underlying community structure, meaning $y_i = 0$ and $z_i = d_i$ if i is an outlier.

Once nodes are assigned to communities and their degrees are split, the edges of each community are then independently generated by the configuration model on the corresponding community degree sequences. After that, the background graph is generated by the configuration model on the degree sequence $(z_i, i \in [n])$. The final **ABCD+o**^{2} model, after an additional clean-up phase handling possible self-loops and duplicate edges, is the union of the community graphs and the background graph.

2.5 The ABCD+o^2 Construction

The following 6-phase construction process generates the **ABCD+o**^{2} synthetic networks.

Phase 1: Creating the Degree Distribution. This phase is the same as in the original **ABCD** model and its generalization, **ABCD+o**. The degree distribution of **ABCD+o**^{2} can be injected into the model as an input. However, by default, it is a distribution that satisfies (a) a power-law with parameter γ, (b) a minimum value of at least δ, and (c) a maximum value of at most Δ.

To achieve the desired degree sequence, degrees are sampled i.i.d. from the distribution $\mathcal{P}(\gamma, \delta, \Delta)$. Let $\mathbf{d}_n = (d_i, i \in [n])$ be the generated degree sequence of G_n with $d_1 \geq \cdots \geq d_n$. Finally, to ensure that $\sum_{i \in [n]} d_i$ is even, we decrease d_1 by 1 if necessary; we relabel as needed to ensure that $d_1 \geq d_2 \geq \cdots \geq d_n$.

Phase 2: Assigning Nodes as Outliers. This phase is also the same as in the **ABCD+o** model. As mentioned in the big picture summary, the neighbours of outliers will be sampled from the entire graph, ignoring the underlying community structure. It feels that this part is straightforward, but one potential problem might occur when ξ is close to zero. In the extreme case when $\xi = 0$, only outliers have a non-zero degree in the background graph. In order to make sure that there exists a simple graph that satisfies the required degree distribution, in such extreme situations all outliers must have degrees smaller than s_0.

To prepare for potential problems, the following procedure is proposed in the **ABCD+o** model, which we also keep here. We have that $\ell = \sum_{i \in [n]} \min(1, \xi d_i)$ is a lower bound for the expected number of nodes that will have a non-zero degree in the background graph. Moreover, since outliers have all neighbours in the background graph, there must be at least s_0 nodes of positive degree in the background graph. Assuming that outliers are selected uniformly at random, we expect $\ell + (n - \ell)(s_0/n)$ nodes of positive degree in the background graph. (In fact, since there is a slight bias toward selecting small degree nodes for outliers, we expect slightly more nodes of positive degree in the background graph, which is good.) We introduce the following constraint: a node i of degree d_i can become an outlier if

$$d_i \leq \ell + s_0 - \ell s_0/n - 1. \qquad (2)$$

Finally, s_0 nodes satisfying (2) are selected uniformly at random to become outliers.

Phase 3: Creating Overlapping Communities. By the end of Phase 2, we have a degree sequence $(d_i, i \in [n])$ and an assignment of outliers and non-outliers. We next assign communities to the **ABCD+o^2** model. It is important to keep in mind that overlapping communities are created in this phase but we do not assign specific nodes to these communities just yet. This assignment process will be handled in the next phase, Phase 4. To make sure there is no confusion, in this phase we will be referring to overlapping sets of *elements (not nodes!)* and in the next phase we will match non-outlier nodes with elements of these sets.

The communities we create here will overlap, provided that $\eta > 1$. There are $\hat{n} = n - s_0$ elements that will eventually be matched with non-outliers and, at the end of this phase, we would like them to belong to $\eta \geq 1$ communities, on average. To achieve this goal and to be compatible with the original **ABCD** model, each non-outlier will belong to a single **primary** community and possibly some **secondary** communities. In particular, this ensures that primary memberships form a partition of non-outlier nodes. We will first generate this partition and then grow each part by a factor of η so that the collective size of all communities is equal to $\eta \hat{n} = \eta(n - s_0)$ in expectation.

Similar to the degree distribution, the distribution of community sizes $(s_j, j \in [L])$ will satisfy (a) a power-law with parameter β, (b) a minimum value of s, and (c) a maximum value of S. Hence, the distribution of primary communities

($\hat{s}_j, j \in [L]$) needs to satisfy power-law with the same parameter β but with a minimum value of $\hat{s} = \lceil s/\eta \rceil$ and a maximum value of $\hat{S} = \lfloor S/\eta \rfloor$. In addition, we require $\sum_{j \in [L]} \hat{s}_j = \hat{n}$. To satisfy both requirements, communities are generated with sizes determined independently by the distribution $\mathcal{P}\left(\beta, \hat{s}, \hat{S}\right)$ until their collective size is at least $\hat{n}$. If, at this point, the sum is $\hat{n} + a$ with $a > 0$ then we perform one of two actions: if the last added community has size at least $a + s$, then we reduce its size by a. Otherwise (that is, if its size is $c < a + s$), then we delete this community, select $c - a$ old communities and increase each of their sizes by 1.

Let L be the random variable counting the number of communities (ignoring the auxiliary "community" C_0 consisting of outliers). Each primary community of size $\hat{s}_j$ will grow to size $s_j = \lfloor \eta \hat{s}_j \rceil$. For $a \in \mathbb{Z}$ and $b \in [0, 1)$ define the random variable $\lfloor a + b \rceil$ as

$$\lfloor a + b \rceil = \begin{cases} a & \text{with probability } 1 - b, \text{ and} \\ a + 1 & \text{with probability } b. \end{cases}$$

(Note that $\mathbb{E}[\lfloor a + b \rceil] = a(1 - b) + (a + 1)b = a + b$.) As a result,

$$\mathbb{E}\left[\sum_{j \in [L]} s_j\right] = \sum_{j \in [L]} \mathbb{E}[s_j] = \eta \sum_{j \in [L]} \hat{s}_j = \eta \hat{n} = \eta(n - s_0),$$

as desired.

For communities to overlap in a natural way, we first create a hidden **reference** layer that will guide the process of assigning elements to specific, overlapping communities. One may think of this auxiliary layer as various latent properties of objects associated with nodes (such as people's age, education, geographic location, beliefs, etc.) shaping communities (such as communities in social media). In this reference layer, each of the $\hat{n}$ elements is assigned a random vector in $\mathbb{R}^2$ that is taken independently and uniformly at random from the ball of radius 1 centred at $\mathbf{0} = (0, 0)$.

Recall that the sequence $(\hat{s}_j, j \in L)$ of primary community sizes is already generated. Let R be the set of $\hat{n}$ elements. We assign these elements to communities, dealing with one primary community at a time, in a random order. When a primary community $\hat{C}_j$ is about to be formed, we first select an element from R that is at the furthest distance from the center $\mathbf{0}$ (in the reference layer). This element, together with its $\hat{s}_j - 1$ nearest neighbours in R, are put to $\hat{C}_j$. We remove $\hat{C}_j$ from R and move on to generating the next primary community. Once all elements are assigned to primary communities, we get a partition; each element belongs to a single community which we call its **primary** community.

Now it is time to grow each primary community of size $\hat{s}_j$ so that its final size is s_j. We can grow communities in any order, as the order will not matter. As before, let R be the set of all $\hat{n}$ elements (in the reference layer). For a given primary community $\hat{C}_j$ of size $\hat{s}_j$, we first compute the center of mass of elements assigned to this primary community, $\mathbf{x}_j \in \mathbb{R}^2$. Then, we investigate elements in R in the order of increasing distances from $\mathbf{x}_j$. If some element $v \in R$ is not already a primary member of this community, we assign this community to v as its **secondary** community. We stop the procedure once the number of members of this community (both primary and secondary) is exactly s_j. We will use $C_j \supseteq \hat{C}_j$ to denote this community, $|C_j| = s_j \geq \hat{s}_j = |\hat{C}_j|$.

Note that each element belongs to exactly one primary community but can be part of many (or none) secondary communities. In Fig. 1 we show an example of the reference layer on $\hat{n} = 150$ elements and three communities. Each of the three primary communities in this example consists of 50 elements before growing by a factor of $\eta = 1.5$, attracting an additional 25 elements as its secondary members.

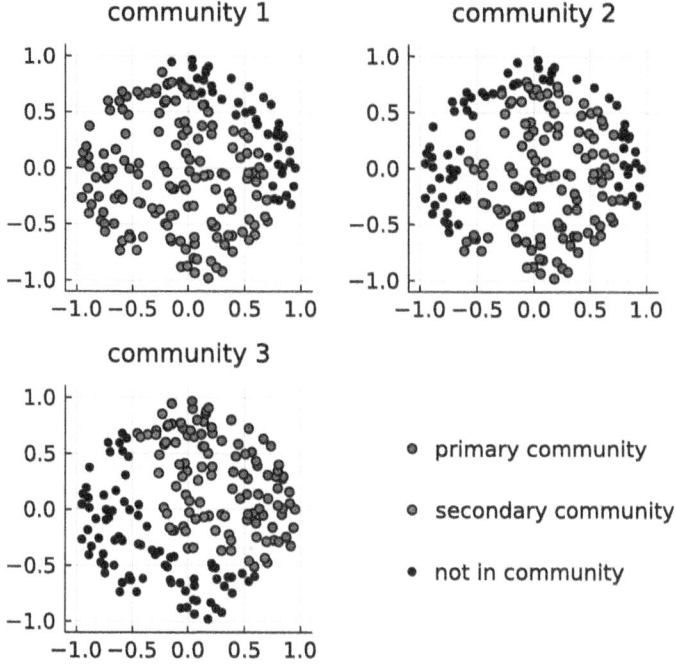

Fig. 1. Example of the reference layer on $\hat{n} = 150$ elements consisting of 3 overlapping communities with equal sizes and $\eta = 1.5$.

Phase 4: Assigning Degrees to Nodes. At this point in the construction of **ABCD+o**^{2} we have a degree sequence $(d_i, i \in [n])$, an assignment of outliers to degrees in the sequence, and a collection of overlapping communities containing "elements", each element having a primary community and some number of secondary communities. Let $\hat{\mathbf{d}}_{\hat{n}}$ be the subsequence (of length $\hat{n}$) of $(d_i, i \in [n])$ corresponding to the non-outliers. We are now ready to assign degrees in $\hat{\mathbf{d}}_{\hat{n}}$ to the community elements. Each element $j \in [\hat{n}]$, that will eventually get node i of degree d_i assigned, is expected to have ξd_i neighbours in the background graph and the remaining $(1-\xi)d_i$ neighbours split evenly between $\eta_j \geq 1$ communities. Note that, although each element has a distinct primary community, its degree will be split with no preference given to said primary community.

Similarly to the potential problem with outliers, we need to make sure that non-outliers of a large degree do not join small communities. Although, for a node $j \in [\hat{n}]$, we know the expected fraction of j's neighbours belonging to community C_k, $k \in [L]$, this is in fact a lower bound as some neighbours of j from the background graph might be in C_k by chance. To make enough room in the community graph, a small correction (typically negligible in practice) is introduced in both **ABCD** and **ABCD+o** that is guided by the parameter ϕ (typically ϕ is very close to 1). For consistency, we keep it in the **ABCD+o**^{2} model as well.

We iteratively assign degrees to elements as follows. Recall that the degree sequence $\hat{\mathbf{d}}_{\hat{n}}$ is sorted with $\hat{d}_1$ being the maximum degree. Starting with $i = 1$, let U_i be the collection of unassigned elements at step i. At step i, choose an element j uniformly at random from the set of elements in U_i that satisfy

$$\hat{d}_i \leq \frac{\eta_j}{1-\xi\phi} \cdot \min\left\{|C_k| - 1 : j \in C_k\right\}, \tag{3}$$

where η_j is the number of communities element j belongs to and

$$\phi = 1 - \sum_{k \in [L]} \left(\frac{\hat{s}_k}{\hat{n}}\right)^2 \frac{\hat{n}\xi}{\hat{n}\xi + s_0},$$

and assign this element j to the ith node in the subsequence $\hat{\mathbf{d}}_{\hat{n}}$ that is of degree $\hat{d}_i$; we have that $U_{i+1} = U_i \setminus \{j\}$.

Recall that, eventually, degree d_i of node i belonging to η_i communities will be split into the background degree (approximately ξd_i) and the community degree that will be further split into η_i parts (approximately $(1-\xi)d_i/\eta_i$ each). This explains the condition (3): $(1-\xi)d_i/\eta_i$ has to be smaller than the smallest community i is part of. Indeed, we bound the degrees assignable to element j in the community C_k to ensure that there are enough elements in $C_k \setminus \{j\}$ for j to pair with, preventing guaranteed self-loops or guaranteed multi-edges during the next phase of the construction. Element j could possibly belong to multiple communities, but the bottleneck is clearly with the smallest one that is of size $\min\{|C_k| : j \in C_k\}$. This strategy guarantees that the assignment is selected uniformly at random from the set of all admissible assignments. The details are quite involved and not overly important for our present discussion. Thus, we point the reader to [13, 14, 16] for a full explanation of the assignment process.

Phase 5: Creating Edges. At this point there are n nodes with labels from $[n]$; $\hat{n} = n - s_0$ of them are non-outliers and the remaining ones are outliers. There is also a family of overlapping communities with each non-outlier node $i \in [n]$ belonging to $\eta_i \geq 1$ communities. Finally, each node $i \in [n]$ (either outlier or non-outlier) is assigned a degree d_i which we interpret as a set of d_i unpaired half-edges. The last step is to form the edges.

For each non-outlier $i \in [n]$ we split its d_i half-edges into *community* half-edges and *background* half-edges. To this end, define $Y_i := \lfloor (1-\xi)d_i \rceil$ and $Z_i := d_i - Y_i$ (note that Y_i and Z_i are random variables with $\mathbb{E}[Y_i] = (1-\xi)d_i$ and $\mathbb{E}[Z_i] = \xi d_i$) and, for all non-outliers $i \in [n]$, split the d_i half-edges of i into Y_i community half-edges and Z_i background half-edges. Community half-edges are further split into η_i communities non-outlier node i belongs to, as evenly as possible. Specifically, for the communities containing node i, $Y_i - \eta_i \lfloor Y_i/\eta_i \rfloor$ communities (chosen randomly) each receive $\lfloor Y_i/\eta_i \rfloor + 1$ half-edges and the remaining communities each receive $\lfloor Y_i/\eta_i \rfloor$ half-edges. On the other hand, if $i \in [n]$ is an outlier then we set $Z_i = d_i$.

Once the assignment of degrees is complete, for each $j \in [L]$, we independently construct the *community graph* $G_{n,j}$ as per the configuration model on node set C_j and the corresponding degree sequence. In the event that the sum of degrees in a community is odd, we pick a maximum degree node i in said community and decrease its community degree by one while increasing its background graph degree by one. Finally, construct the *background graph* $G_{n,0}$ as per the configuration model on node set $[n]$ and degree sequence $(Z_i, i \in [n])$. Let $G_n = \bigcup_{0 \le j \le n} G_{n,j}$ be the union of all graphs generated in this phase.

Phase 6: Rewiring Self-loops and Multi-edges. Note that, although we are calling $G_{n,0}, G_{n,1}, \ldots, G_{n,L}$ *graphs*, they are in fact *multi-graphs* at the end of phase 5. To ensure that G_n is simple, we perform a series of *rewirings* in G_n. A rewiring takes two edges as input, splits them into four half-edges, and creates two new edges distinct from the input. We first rewire each community graph $G_{n,j}$ ($j \in [L]$), and the background graph $G_{n,0}$, independently as follows.

1. For each edge $e \in E(G_{n,j})$ that is a loop, we add e to a *recycle* list that is assigned to $G_{n,j}$. Similarly, if $e \in E(G_{n,j})$ contributes to a multi-edge, we put all but one copies of this edge to the *recycle* list.
2. We shuffle the *recycle* list and, for each edge e in the list, we choose another edge e' uniformly from $E(G_{n,j}) \setminus \{e\}$ (not necessarily in the list) and attempt to rewire these two edges. We save the result only if the rewiring does not lead to any further self-loops or multi-edges, otherwise we give up. In either case, we then move to the next edge in the *recycle* list.
3. After we attempt to rewire every edge in the *recycle* list, we check to see if the new *recycle* list is smaller. If yes, we repeat step 2 with the new list. If no, we give up and move all of the "bad" edges from the community graph to a collective *global recycle* list.

As a result, after ignoring edges in the *global recycle* list, all community graphs are simple and the background graph is simple. However, as is the case in the original **ABCD** model, an edge in the background graph can form a multi-edge with an edge in a community graph. Another problem that might occur, specific to **ABCD+o**^{2} model, is that an edge from one community can form a multi-edge with an edge from a different but overlapping community. All of these problematic edges are added to the *global recycle* list. We merge all community graphs with the background graph. Finally, the *global recycle* list is transformed into a list of half-edges and new edges are created from it. We follow the same procedure as for the community graphs. However, we do not "give up" recycling and follow the process until all required edges are created. As the background graph is sparse, this final rewiring is very fast in practice.

3 Properties of the ABCD+o² Model

In this section, we present the results of some experiments highlighting properties of the **ABCD+o²** model.

Degree Distribution and Community Size Distribution. In the first experiment, we generate **ABCD+o²** graphs with three degree sequences and with $n = 100{,}000$. The minimum and the maximum degrees are fixed to be $\delta = 5$, $\Delta = 316 \approx \sqrt{n}$, but the power-law exponents vary: $\gamma \in \{2.2, 2.5, 2.8\}$. For a given integer k, let $f(k)$ be the experimental cumulative degree distribution, that is, $f(k)$ is the fraction of nodes of degree at least k. For a given set of parameters, the theoretical cumulative degree distribution is given by (1). We show that the experimental degree distributions are very close to the desired, theoretical, ones—see Fig. 2 (Left) for the cumulative degree distributions of the three sequences.

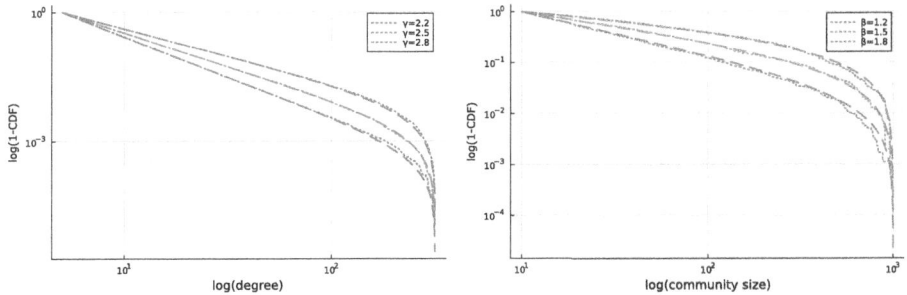

Fig. 2. Left: Empirical (dots) and theoretical (dashes) cumulative degree distributions for three degree sequences: $n = 100\,000$, $\delta = 5$, $\Delta = 316$. Right: Empirical (dots) and theoretical (dashes) cumulative community sizes distributions for three sizes sequences: $n = 100\,000$, $s = 10$, $S = 1\,000$.

Similarly, we generate three sequences of community sizes with $n = 100{,}000$. The minimum and the maximum community size is fixed to be $s = 10$ and $S = 1000$ for the three sequences, but the power-law exponents vary: $\beta \in \{1.2, 1.5, 1.8\}$. We show that the experimental sequences are also very close to the desired ones—see Fig. 2 (Right) for the cumulative community sizes distributions.

Overlapping of Communities. The **ABCD+o²** model generates random graphs in which non-outlier nodes belong to η communities, on average. Let ρ_k be the fraction of non-outliers that belong to exactly k communities. The sequence $(\rho_k)_{k \geq 1}$ depends on the structure of the underlying hidden layer. To show one example, we generated five **ABCD+o²** graphs with varying overlap parameter $\eta \in \{1, 1.5, 2, 2.5, 3\}$. Each of these graphs consists of $n = 10{,}000$ nodes, including $s_0 = 250$ outliers, node degrees in range $[10, 100]$ with power law exponent $\gamma = 2.5$, and community sizes in range $[50, 1170]$ with power law exponent $\beta = 1.5$. The corresponding sequences $(\rho_k)_{k \geq 1}$ are presented in Fig. 3 (left). In the case when $\eta = 3$, we also give the number and size of the non-empty overlaps between 2, 3 or 4 communities—see Fig. 3 (right).

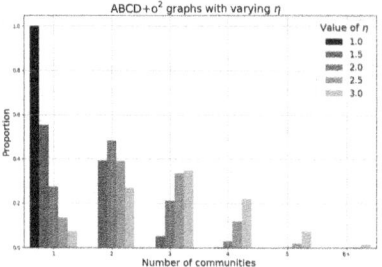 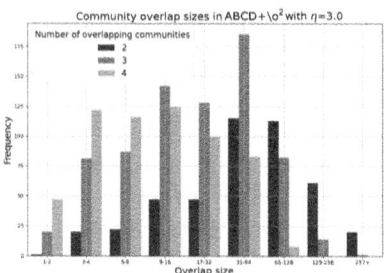

Fig. 3. Distribution of the number of community memberships for non-outlier nodes on **ABCD+o²** graphs with $n = 10{,}000$ and varying η (left), and distribution of overlap sizes in the non-empty intersections between 2, 3 or 4 communities (right).

Community Association Strength. The **ABCD+o²** model aims to generate graphs in which nodes are much more associated with the communities they belong to than with other communities. The next experiment suggests that this goal is achieved.

In the coming experiment, we use three measures of community association strength: Internal Edge Fraction (IEF), Normalized Internal Edge Fraction (NIEF), and P-score (P). For node i and community C, IEF(i, C) is the fraction of i's edges with the other end-point in C, NIEF$(i, C) = $ IEF$(i, C) - \mathbb{E}\left[\text{IEF}(i, C)\right]$, where expectation is taken with respect to the Chung-Lu null model, and $P(i, C)$ is based on the classic p-value test, i.e., based on the probability that the IEF(i, C) score was achieved randomly (again, using Chung-Lu as a null model). We point the interested reader to [2] for a more thorough discussion of these three measures.

For the experiment, we generate two **ABCD+o²** graphs, one with a low level of noise ($\xi = 0.35$) and the other with a high level of noise ($\xi = 0.65$). For both cases, we computed the following. For each of the three community association strength measures, and for each value of $K \in \mathbb{N}$, we investigate all nodes that belong to at least K communities and check what fraction of them have their K'th top ranked community (with respect to a given association strength) align with one of their ground-truth communities. The results are presented in Fig. 4. The results suggest that each of the measures can accurately predict 1 or 2 communities a node is a member of, and with a low noise parameter, the prediction accuracy remains high as the number of communities increases.

4 Benchmarking Community Detection Algorithms

The main purpose of having synthetic models with ground-truth community structure is to test, tune, and benchmark community detection algorithms. To showcase **ABCD+o^2** in this light, we use the model to evaluate the performance of five community detection algorithms. The algorithms are as follows.

Leiden [23]: a greedy algorithm that attempts to optimize the modularity function. Note that this algorithm returns a partition, and we use it merely as a benchmark to compare with algorithms that attempt to find overlapping communities.

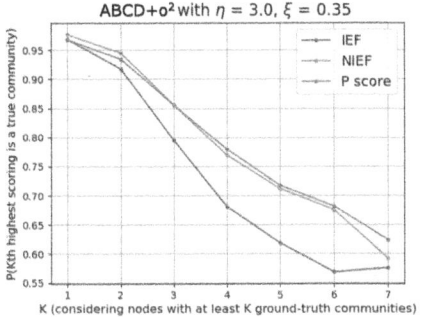 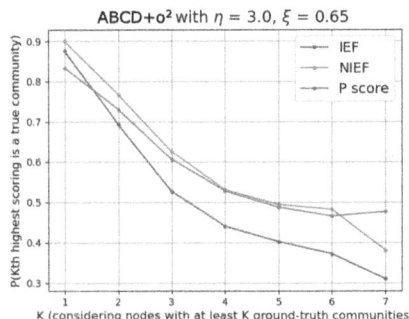

Fig. 4. The top-K CAS scores for nodes with K community memberships or more. For each score and each K, we show the proportion of true communities respectively for low noise **ABCD+o^2** graphs (left) and high noise (right).

Clique Percolation [6]: an algorithm, based on a positive integer k, that finds all k-cliques and declares two such cliques adjacent if they share $k-1$ nodes. Then, the connected collections of cliques yield a collection of overlapping communities. In our experiment, we choose $k = 3$.

Edge Clustering [18]: an edge-partitioning algorithm that translates to overlapping clusters of nodes. Here, pairs of edges are measured based on similarity of neighbourhoods, and these similarity measures dictate the order in which edge-communities merge, starting from each edge in its own community. As edge-communities merge, the modularity is tracked on the line-graph, and the maximum modularity attained yields the edge-communities, which in turn yields overlapping node-communities.

Ego-Split [7]: a method which finds overlapping clusters in a graph G by applying a partitioning algorithm such as Leiden to an auxiliary graph G' and then mapping the resulting partition onto G. The auxiliary graph G' is constructed from G by creating multiple copies, or "egos", of each node based on its neighbourhood.

Ego-Split+CAS: the same algorithm as Ego-Split, but with a post-processing step that re-assigns nodes to communities based on the NIEF measure.

This is by no means an exhaustive list of community detection algorithms. We wish only to showcase the usefulness of **ABCD+o^2** in comparing the quality of detection algorithms. The measure we use to determine the quality of a collection of communities is the overlapping Normalized Mutual Information (oNMI) measure: a similarity measure for two collections of subsets $\mathcal{X}, \mathcal{Y}$ of a set S [21].

The parameters of **ABCD+o^2** with the most influence on the quality of detection algorithms are ξ (the level of noise) and η (the average number of communities a non-outlier is part of). Thus, we perform two versions of this experiment, one which varies $\xi \in \{0.15, 0.25, \ldots, 0.65\}$ and fixes $\eta = 2$, and the other which varies $\eta \in \{1, 1.5, 2, 2.5, 3\}$ and fixes $\xi = 0.15$. Figure 5 presents the results of the experiment. We see that Ego-Split+CAS performs the best overall, except when $\eta = 1$ in which case Leiden performs better. We also see a general trend of all algorithms performing worse as the graph gets noisier, either by increasing ξ or η. From numerous and varying tests, we have found in general that increasing η is far more damming to detection algorithms than increasing ξ.

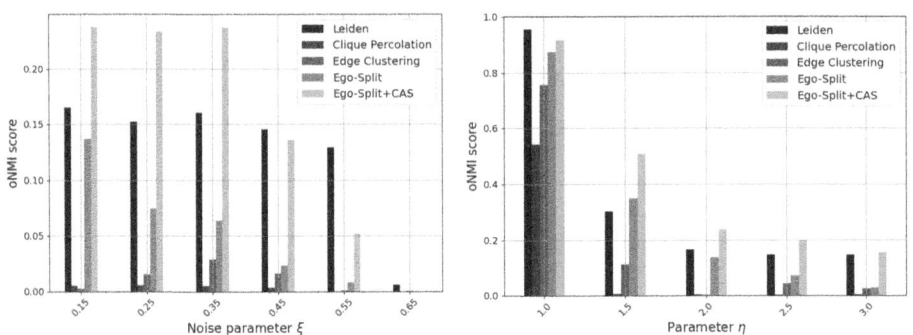

Fig. 5. The quality of five clustering algorithms on **ABCD+o^2** graphs with 10,000 nodes including 250 outlier nodes. In the left plot, we fix $\eta = 2.0$ and vary ξ. In the right plot, we fix $\xi = 0.15$ and vary η.

5 Conclusion

We presented **ABCD+o^2**: a generalization of the **ABCD+o** model that allows for overlapping communities. We then tested properties of this new model, emphasizing properties based on the new overlap parameter η. Finally, we showcased the model's ability to benchmark community detection algorithms and compare their quality.

This paper acts as a first step in our study of the **ABCD**+**o**^{2} model. In future work, we will study more properties of the model and compare our findings with (i) real networks containing overlapping ground-truth communities, and (ii) the overlapping **LFR** model. In particular, we believe that the nature of community overlap (based on the hidden, geometric reference layer) is more natural and realistic in **ABCD**+**o**^{2} than in **LFR** and we will explore this conjecture further.

We are interested in theoretical results of the **ABCD**+**o**^{2} model that generalize results of the **ABCD** and **ABCD**+**o** models. In [13] the modularity was studied and it was found that the maximum modularity came from the ground truth communities until a certain level of noise, after which a higher modularity could be attained. A similar behaviour should be seen with **ABCD**+**o**^{2} and its overlap parameter η. Additionally, in [3] it was shown that **ABCD** graphs exhibit self-similar behaviour, namely, the degree distributions of communities are asymptotically the same as the degree distribution of the whole graph (up to an appropriate normalization). We suspect that this self-similar property persists in **ABCD**+**o**^{2}.

Finally, we are interested in modifying the **ABCD**+**o**^{2} model in various ways. On the one hand, the underlying geometry is the key ingredient in forming overlaps between communities, and changing this geometry will surely change the behaviour of the overlap. Moreover, certain metric spaces may yield **ABCD**+**o**^{2} graphs with more realistic properties. On the other hand, the assignment of degrees to nodes can be tweaked, say, to bias large degrees towards nodes that are members of a large number of communities. In [26] it was shown that real networks tend to have a higher density of edges in the intersections of communities than in the communities themselves. We hope that by tweaking the degree-to-node assignment process in **ABCD**+**o**^{2} we can find this same density result.

References

1. Aref, S., Chheda, H., Mostajabdaveh, M.: The Bayan algorithm: detecting communities in networks through exact and approximate optimization of modularity. arXiv preprint arXiv:2209.04562 (2022)
2. Barrett, J., DeWolfe, R., Kamiński, B., Prałat, P., Smith, A., Théberge, F.: Improving community detection via community association strength scores (2025). https://arxiv.org/abs/2501.17817
3. Barrett, J., Kamiński, B., Prałat, P., Théberge, F.: Self-similarity of communities of the ABCD model. Theor. Comput. Sci. **1026**, 115012 (2025)
4. Bender, E.A., Rodney Canfield, E.: The asymptotic number of labeled graphs with given degree sequences. J. Comb. Theory Ser. A **24**(3), 296–307 (1978)
5. Bollobás, B.: A probabilistic proof of an asymptotic formula for the number of labelled regular graphs. Eur. J. Comb. **1**(4), 311–316 (1980)
6. Derényi, I., Palla, G., Vicsek, T.: Clique percolation in random networks. Phys. Rev. Lett. **94**(16), 160202 (2005)
7. Epasto, A., Lattanzi, S, Leme, R.P.: Ego-splitting framework: From non-overlapping to overlapping clusters. In: Proceedings of the 23rd ACM SIGKDD

International Conference on Knowledge Discovery and Data Mining, pp. 145–154 (2017)
8. Fortunato, S.: Community detection in graphs. Phys. Rep. **486**(3–5), 75–174 (2010)
9. Fortunato, S., Barthelemy, M.: Resolution limit in community detection. Proc. Natl. Acad. Sci. **104**(1), 36–41 (2007)
10. Girvan, M., Newman, M.: Community structure in social and biological networks. Proc. Natl. Acad. Sci. **99**(12), 7821–7826 (2002)
11. Gregory, S.: Fuzzy overlapping communities in networks. J. Stat. Mech: Theory Exp. **2011**(02), P02017 (2011)
12. Kamiński, B., Olczak, T., Pankratz, B., Prałat, P., Théberge, F.: Properties and performance of the ABCDe random graph model with community structure. Big Data Res. **30**, 100348 (2022)
13. Kamiński, B., Pankratz, B., Prałat, P., Théberge, F.: Modularity of the ABCD random graph model with community structure. J. Complex Netw. **10**(6), cnac050 (2022)
14. Kamiński, B., Prałat, P., Théberge, F.: Artificial benchmark for community detection (ABCD)-fast random graph model with community structure. Netw. Sci. 1–26 (2021)
15. Kamiński, B., Prałat, P., Théberge, F.: Mining Complex Networks. Chapman and Hall/CRC (2021)
16. Kamiński, B., Prałat, P., Théberge, F.: Artificial benchmark for community detection with outliers (ABCD+o). Appl. Netw. Sci. **8**(1), 25 (2023)
17. Kamiński, B., Prałat, P., Théberge, F.: Hypergraph artificial benchmark for community detection (h–ABCD). J. Complex Netw. **11**(4), cnad028 (2023)
18. Kim, P., Kim, S.: Detecting overlapping and hierarchical communities in complex network using interaction-based edge clustering. Physica A: Stat. Mech. Appl. **417**, 46–56 (2015). . https://doi.org/10.1016/j.physa.2014.09.035. https://www.sciencedirect.com/science/article/pii/S0378437114007936
19. Lancichinetti, A., Fortunato, S.: Benchmarks for testing community detection algorithms on directed and weighted graphs with overlapping communities. Phys. Rev. E **80**(1), 016118 (2009)
20. Lancichinetti, A., Fortunato, S., Radicchi, F.: Benchmark graphs for testing community detection algorithms. Phys. Rev. E **78**(4), 046110 (2008)
21. McDaid, A.F., Greene, D., Hurley, N.: Normalized mutual information to evaluate overlapping community finding algorithms. arXiv preprint arXiv:1110.2515 (2011)
22. Palla, G., Derényi, I., Farkas, I., Vicsek, T.: Uncovering the overlapping community structure of complex networks in nature and society. Nature **435**(7043), 814–818 (2005)
23. Traag, V., Waltman, L., van Eck, N.J.: From Louvain to Leiden: guaranteeing well-connected communities. Sci. Rep. **9**, 5233 (2019). https://doi.org/10.1038/s41598-019-41695-z
24. Wormald, N.C.: Generating random regular graphs. J. Algorithms **5**(2), 247–280 (1984)
25. Wormald, N.C., et al.: Models of random regular graphs. Lond. Math. Soc. Lecture Note Series 239–298 (1999)
26. Yang, J., Leskovec, J.: Overlapping communities explain core-periphery organization of networks. Proc. IEEE **102**(12), 1892–1902 (2014). https://doi.org/10.1109/JPROC.2014.2364018

A Graph Network Approach to Disinformation Detection in Social Media

Milita Songailaitė[1,2(✉)], Justina Mandravickaitė[1,2], Veronika Bryskina[1,2], Maksym Bondar[1,2], and Tomas Krilavičius[1,2]

[1] Faculty of Informatics, Vytautas Magnus University, Kaunas, Lithuania
milita.songailaite@vdu.lt
[2] Center for Applied Research and Development, Kaunas, Lithuania
info@card-ai.eu
https://if.vdu.lt/en/

Abstract. Our study addresses the challenge of processing vast amounts of unstructured data by extracting and organizing key information via knowledge graphs. Focusing on disinformation detection, we analysed over 100,000 messages from nine Russian Telegram channels. We constructed knowledge graphs applying two triple extraction methods: Subject-Verb-Object (SVO) and Entity-Relation-Entity (ERE). We identified and studied disinformation cases by comparing Telegram data graphs with graphs of verified cases from the EUvsDisinfo database using Graph Kernels. Results revealed disinformation across all analysed channels, particularly regarding the downing of Flight MH17, claims about Western assistance, and narratives on Ukraine's occupied regions. Our results demonstrated that the Shortest Path Graph Kernel and Subgraph Matching methods were the most effective for detecting disinformation as they accurately identified graphs containing substantial disinformation, highlighting the potential of knowledge graphs in large-scale media monitoring.

Keywords: Disinformation detection · Graph kernels · Information extraction · Similarity distances · Social media

1 Introduction

This paper presents experiments that apply graph-based methods to detect and analyze disinformation, focusing on disinformation in Russian Telegram. Using data from 9 Telegram channels and documented disinformation cases from the EUvsDisinfo database, we extracted relational triples and used them to construct separate knowledge graphs (KGs) for each dataset. Subsequently, we applied graph kernel methods to compare these KGs and identify disinformation cases and trends.

Monitoring social and traditional media involves processing vast amounts of unstructured data which is characterized by inconsistent format, diverse granularity, and linguistic as well as cultural variability [16]. Addressing these challenges requires extracting key information to organize the data into manageable formats for effective analysis and decision-making. Therefore, data reduction is a crucial step in this process. Advanced techniques, including natural language processing (NLP), machine learning, and information extraction, are employed to identify and summarize the most relevant information [12].

Also, structured data representations, such as knowledge graphs [6], relational databases [39], or predefined ontological schemas [13], are essential for transforming unstructured content. These representations simplify querying, improve visualization and integration into analytical processes. By converting unstructured data into structured formats, the challenges of the volume and complexity of media content can be mitigated, increasing the efficiency of media monitoring [32]. This approach enables the identification of key trends, detection of disinformation [1], and generation of important insights, thus contributing to informed decision-making in ever-changing environments. Therefore, graph-based methods are applied in this paper to detect and analyze disinformation, with an emphasis on Russian Telegram content.

The rest of the paper is organized as follows: Sect. 2 introduces related work, Sect. 3 describes the data, Sect. 4 outlines the methods, Sect. 5 presents the results, and Sect. 6 concludes the study.

2 Related Work

Disinformation detection is a non-trivial challenge that requires diverse methodologies to identify and mitigate false or misleading information effectively. Linguistic feature-based methods analyze syntax and other patterns to detect deceptive language via techniques like n-grams, and TF-IDF [21,40]. Deceptive modeling uses NLP and data mining to verify information, such as rumors, through tracking, stance classification, and veracity classification [34]. Also, clustering methods employ graph-based techniques to group content by attributes and detect disinformation through user activity patterns [28]. Meanwhile, predictive modeling applies machine learning methods, e.g., logistic regression, random forests, and SVM, to classify content based on features such as user engagement or temporal trends [10,35]. Additionally, content-based approaches enhance classification results by extracting elements such as source credibility, metadata, and textual features [10].

Furthermore, deep learning (DL) has contributed to advances in disinformation detection significantly. Models like fine-tuned BERT and ensemble methods [18] have shown remarkable results, and architectures such as CNN-LSTM, Bi-LSTM, and LSTM-CNN combinations have been widely used [3] for this task. However, DL models often lack interpretability, so Explainable AI (XAI) techniques have been applied to improve the transparency of model decision-making processes [22].

Moreover, KGs and network analysis provide a promising direction as they enable mapping complex relationships and contextualizing data. KGs have been used in various tasks, such as analysing communication patterns and tracking news narratives [11,30]. Their ability to integrate heterogeneous information also makes them valuable for disinformation analysis [25].

Finally, recent advancements in disinformation detection and analysis include large language models (LLMs) integrated with graph-based reasoning frameworks such as Graph of Thoughts [4], LLM-graph neural network combinations [15], and hypergraph methods [31]. Also, adversarial systems that use reinforcement learning and multi-modal techniques for semantic tracking have been developed to counter disinformation [2,8].

3 Data

3.1 Dataset of Telegram Messages

This dataset consists of messages extracted from the *Telegram* messaging platform[1] and includes 113,770 posts published between January and November 2024 on 9 Russian language channels. These channels were selected based on their reporting on Russia-Ukraine war-related content. Data cleaning involved removing messages that contained fewer than six words, as well as user mentions, tags, and emojis. Initial testing revealed that Russian language models were less accurate than English ones for dependency tagging. Therefore, the messages were translated into English using machine translation and processed using English-specific models. While translation accuracy wasn't perfect (some details, especially slang, were lost in translation), expert analysis showed this approach was of much higher quality than using Russian language models.

To get familiar with the data, an analysis including basic statistics and topic modelling using Latent Dirichlet Allocation (LDA) [17] has been carried out. Statistical analysis revealed that most channels exhibited infrequent posting patterns, except for RIAN, which is a corporate news outlet. Interestingly, channels with high message volume used fewer words per message, while less active channels posted longer messages. After basic statistical analysis, the messages in each channel were clustered into 3 topics using the LDA topic modelling approach. These topics were then categorized into five primary themes (Fig. 1), providing valuable insights into thematic trends and the specific focuses of each channel to direct our further research.

3.2 Dataset of Disinformation Cases

To investigate patterns of disinformation, a dataset was built from the *EuVS-Disinfo* database[2]. This repository, established by the European Union, aims to discover, document, and disprove occurrences of disinformation. The extracted dataset contains 530 disinformation cases that summarize and debunk campaigns from February 24, 2022, to February 20, 2023. Only the cases related to the Russia-Ukraine war are included in this dataset. Pre-processing of the dataset used the same pipeline that was applied to our Telegram dataset.

---

[1] Accessible via: https://web.telegram.org/.
[2] Accessible via: https://euvsdisinfo.eu/misinformation-cases/.

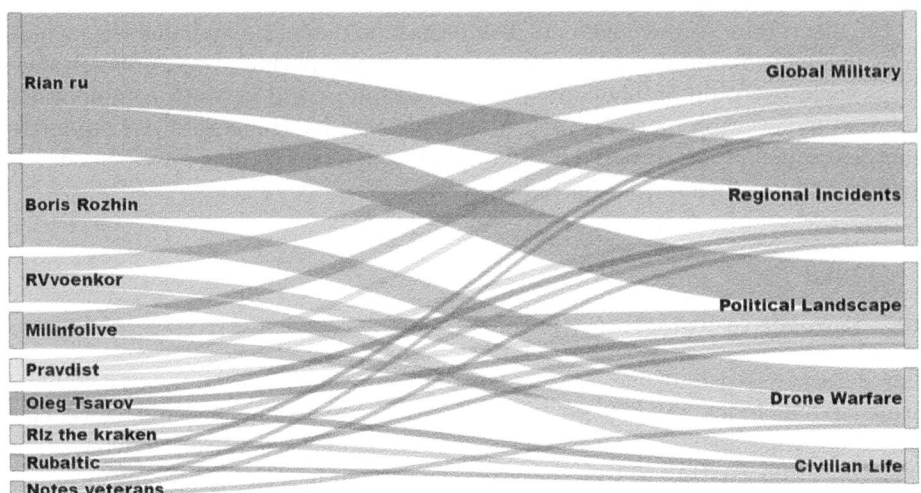

Fig. 1. Topic distribution across Telegram channels

4 Methods

4.1 Extraction of Relational Triples

Triple Extraction via Dependency Parsing. Identifying Subject-Verb-Object (SVO) triples involves extracting subject and object entities along with the verb relation via dependency parsing to ensure grammatical accuracy and a structured representation. [20]. We used $spaCy$[3] dependency parser to analyse sentence structure and identify entities (excluding punctuation). Algorithm 1 illustrates the use of SpaCy dependency parsing to extract entities from unstructured text. It generates entity representations by ignoring punctuation, capturing compound nouns (e.g., "ice cream") and modifiers (e.g., "red apple"), and identifying subjects and objects using *subj* and *obj* tags. The output of this algorithm is a tuple containing a subject and an object for further analysis.

After extracting entities, the next step includes identifying a verb-based relationship in the sentence, which connects the previously extracted Subject and Object. Algorithm 2 uses SpaCy dependency parser to identify the main verb (ROOT) and optionally includes prepositions, agents (passive voice actors), or adjectives that specify the relation. This method first annotates each word with linguistic properties and only then applies a specialized pattern to collect key phrase sequences.

---

[3] Accessible via: https://spacy.io/.

Algorithm 1. Entity Extraction from Sentences

Require: Sentence (string)
Ensure: Pair of extracted entities (subject, object)
1: Initialize $ent1, ent2, dep\_prev\_token, txt\_prev\_token, prefix, modifier$ to empty strings
2: Parse the sentence with NLP tool to get tokens (doc)
3: **for** each token $tokn$ in doc **do**
4:    **if** $tokn.dep\_ ==$ "punct" **then**
5:       Continue to the next token
6:    **end if**
7:    Update $prefix$ and $modifier$ based on $tokn.dep\_$
8:    **if** $tokn.dep\_$ includes "subj" or "obj" **then**
9:       Construct entity string from $modifier, prefix$ and $tokn.text$
10:      Assign entity string to $ent1$ or $ent2$ accordingly
11:      Reset $prefix, modifier$ if detected "subj"
12:    **end if**
13:   Update $dep\_prev\_token, txt\_prev\_token$
14: **end for**
15: **return** $[ent1, ent2]$

Algorithm 2. Relation Extraction from Sentences

Require: Sentence (string)
Ensure: Relation extracted from the sentence (string or None)
1: Initialize the NLP tool to parse the sentence (doc)
2: Create a Matcher instance with the NLP tool's vocabulary (matcher)
3: Apply the matcher to the parsed sentence to find matches (matches)
4: **if** matches are found **then**
5:    Select the last match from the list (h)
6:    Extract the text span from the last match (span)
7:    **return** $span.text$
8: **else**
9:    **return** None
10: **end if**

These SVO triples were used for mapping unstructured text (Telegram messages and descriptions of disinformation cases) to KGs for detecting and analysing disinformation patterns related to targets, narratives, and changes in time.

Triple Extraction via Semantic Relation Extraction. Another approach to triplet extraction involved using semantic relations. We applied REBEL (Relation Extraction By End-to-End Language Generation) [7], a sequence-to-sequence (seq2seq) model based on BART and designed for relational triple extraction. Unlike multi-step pipelines that address Named Entity Recognition (NER) and Relation Classification (RC) individually, REBEL treats relation extraction as text generation. This technique efficiently covers over 200 various relation types while achieving exceptional performance. In this context, REBEL was used to extract Entity-Relation-Entity (ERE) triples from both our datasets (dataset of Telegram messages and dataset of disinformation cases) separately, and these triples then were used to create Knowledge Graphs.

4.2 Construction of Knowledge Graphs

The construction of KGs involves extracting relevant information from unstructured data and assembling it into a graph structure where entities (vertices) are linked by their connections (edges). A knowledge graph $\mathcal{KG}$ can be viewed as a directed graph

$$\mathcal{KG} = (\mathcal{V}, \mathcal{E}, \mathcal{R}),$$

where

- $\mathcal{V} = \{v_1, v_2, ..., v_n\}$ are vertices,
- $\mathcal{R} = \{r_1, r_2, ..., r_k\}$ are relation types,
- $\mathcal{E} = \{(v_i, r_x, v_j) | v_i, v_j \in \mathcal{V}, r_x \in \mathcal{R}\}$ are relation instances (i.e. edges) [23].

In our case, we used two types of relational triples: Entity-Relation-Entity (ERE), which was extracted using REBEL, and Subject-Verb-Object (SVO), which was extracted via dependency parsing. While ERE extraction focuses on finding entities (such as persons, organizations, or concepts) and their semantic relationships regardless of syntax, the SVO technique stresses grammatical roles for recognizing entities and their relations [27]. Therefore, ERE extraction prioritizes higher-level semantic links, which are advantageous for building KGs and evaluating unstructured data across various domains, whereas SVO extraction is strongly linked to sentence structure [38].

4.3 Graph Matching with Graph Kernels

Graph kernels provide a strong mathematical foundation for measuring graph similarity, which is an important topic in many fields like bioinformatics, social network research, and chemistry [5]. Typically, they enable systematic and efficient comparisons among large sets of graphs by capturing their structural similarities.

A graph kernel K is defined as a function $K : \mathcal{G} \times \mathcal{G} \to \mathbb{R}$, where $\mathcal{G}$ is the set of all graphs. The main idea is to project graphs into a high-dimensional feature space and then compute the inner product of their feature representations. Mathematically, for two graphs G_1 and G_2, the kernel is given by:

$$K(G_1, G_2) = \langle \phi(G_1), \phi(G_2) \rangle,$$

where ϕ denotes a feature map that transforms a graph into a vector in some feature space [19].

Many graph kernels fall under the R-convolution framework introduced by Haussler in 1999 [14]. For a graph kernel to be effective and efficient, it must have the following properties [9,26]:

1. **Expressiveness.** It should capture the key similarities and contrasts between the graphs.
2. **Efficient.** It should be able to process huge datasets in polynomial time.
3. **Positively Defined.** This assures that the kernel matrix, which is created by applying the kernel to all pairs of graphs, is symmetric and positive semi-definite.

For comparison and similarity assessment between KG constructed from Telegram messages and KG constructed from EuvsDisinfo disinformation cases, we used five methods:

1. **Shortest Path Kernel** – compares the shortest paths between nodes in our two graphs, focusing on the structural connectivity [19].
2. **Subgraph Matching Kernel** – uses subgraph matching to uncover more complex and precise patterns of the information spread. [24].
3. **Weisfeiler-Lehman Graph Kernel** – assesses more complex patterns of graph similarity via node labeling for graph isomorphism tests [33].
4. **Neighborhood Hash Graph Kernel** – employed hashing to encode neighborhood nodes and then compared these encodings across graphs to discover similarities in local structures [24].
5. **Maximum Common Subgraph measure** – a graph theory concept used to determine the similarity between any two graphs [29] by identifying the largest subgraph shared by any two compared graphs.

5 Results

As part of our disinformation detection strategy, Telegram messages and known disinformation cases were converted into knowledge graphs (KGs), and then compared for similarity. A high similarity between a Telegram KG and a known disinformation KG indicated potential disinformation. After computing similarities, we analyzed: (1) overall disinformation prevalence, (2) frequent disinformation cases, and (3) how disinformation spreads over time. The following subsections expand on these analyses, clarifying disinformation patterns and defining their traits.

5.1 Disinformation Across Analysed Telegram Channels

In the initial matching analysis, each distance measure was averaged. First, all-to-all distances were computed between disinformation-case graphs and Telegram message graphs, with each of the five metrics (Shortest Path Kernel, Subgraph Matching Kernel, Weisfeiler-Lehman Graph Kernel, Neighborhood Hash Graph Kernel, and Maximum Common Subgraph measure) normalized to [0, 1].

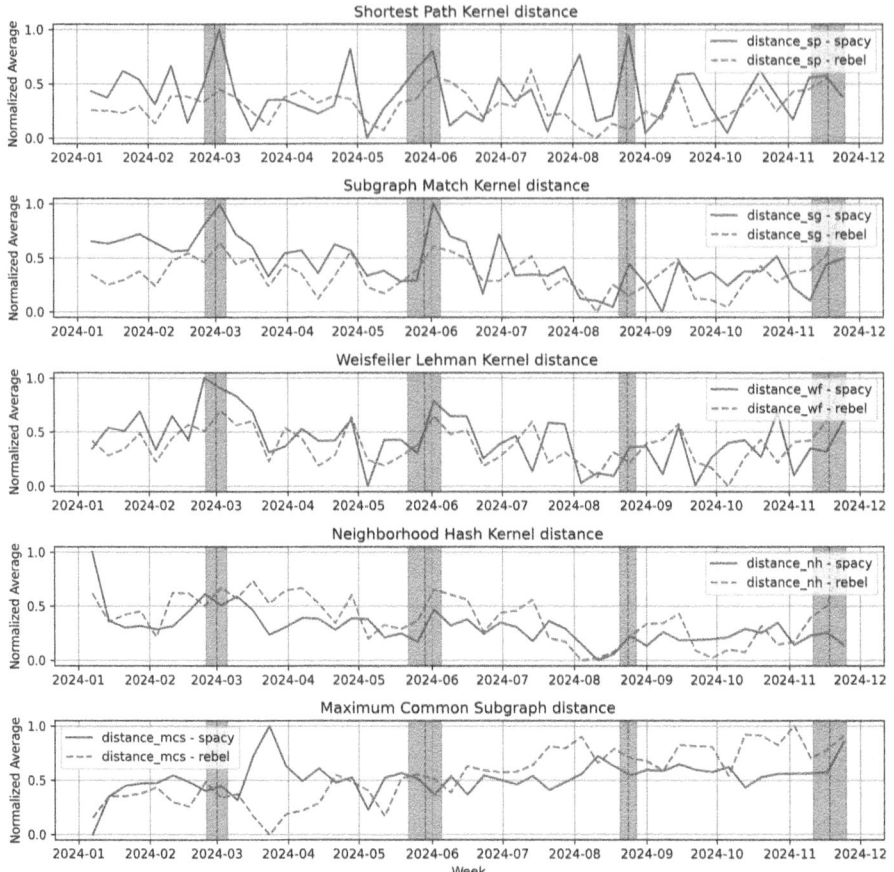

Fig. 2. The averaged and smoothed evaluation of detected disinformation, based on matching graphs from Telegram messages to a set of known disinformation cases. The blue line indicates the results from dependency parsing using SVO triples, while the red dotted line represents the results from ERE triples. The red zones indicate the areas of disinformation spikes, which are discussed in Sect. 5.3. (Color figure online)

Next, only the top 20 most similar disinformation instances per day and channel were used for averaging. These daily averages across five metrics then offered a year-long overview of disinformation dissemination.

A 7-day moving average was used to smooth the time series and remove extraneous variations. Figure 2 shows the results from both dependency parsing and semantic triple approaches. Each of the ten time series (five for each method) underwent an additive decomposition with a seven-day period (Fig. 3), aligning with the earlier smoothing. Although seasonality exists, it may be subtle, and the overall trend shifts significantly over time.

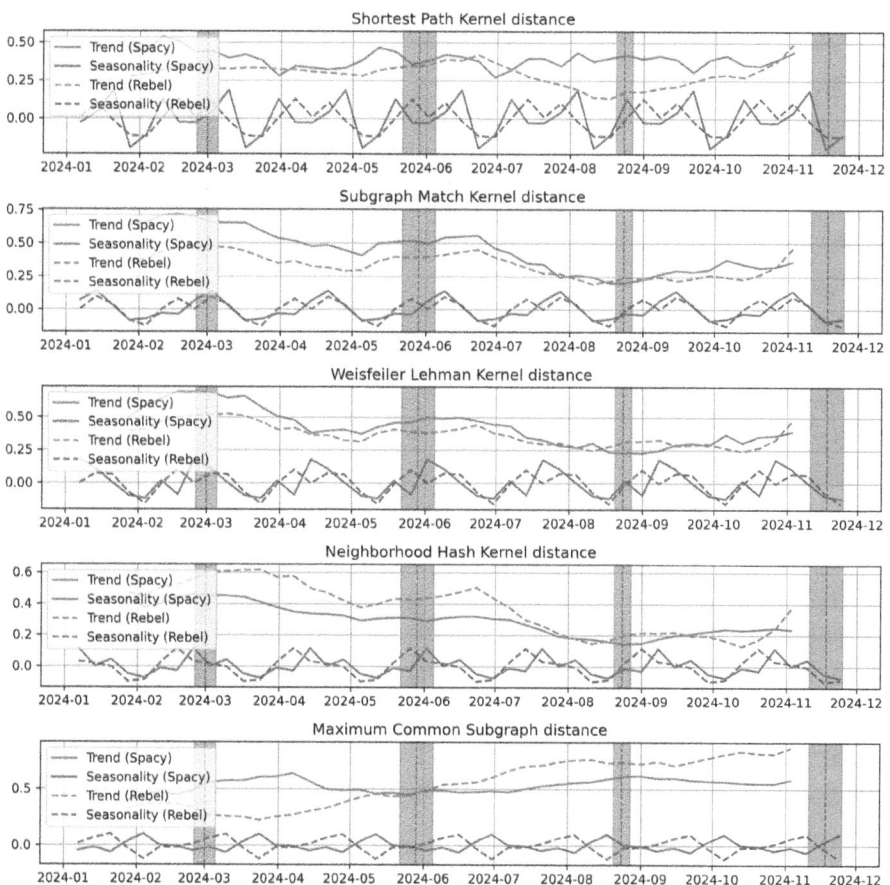

Fig. 3. The trend and seasonality components of all five averaged decomposed time series. The red and blue lines show the trend and seasonality, respectively, derived from dependency parsing with SVO triples, while the dotted orange and purple lines represent the trend and seasonality, respectively, ERE triples. The red zones indicate the areas of disinformation spikes, which are discussed in Sect. 5.3. (Color figure online)

One of the main goals was to measure the disinformation level of each channel, which is shown in Fig. 4. This was accomplished by calculating the disinformation levels for each channel using five graph similarity measures previously discussed. The average of these rankings was then determined (see Fig. 4).

The assessments of disinformation generated by the two triple extraction approaches differed significantly. For instance, using SVO triples, the channel "rian_ru" was found to have the least amount of disinformation, but using ERE triples, it was determined to have the most. In the meantime, several channels displayed patterns that were comparable across both triple extraction approaches. For example, both systems consistently scored "pravdist" and

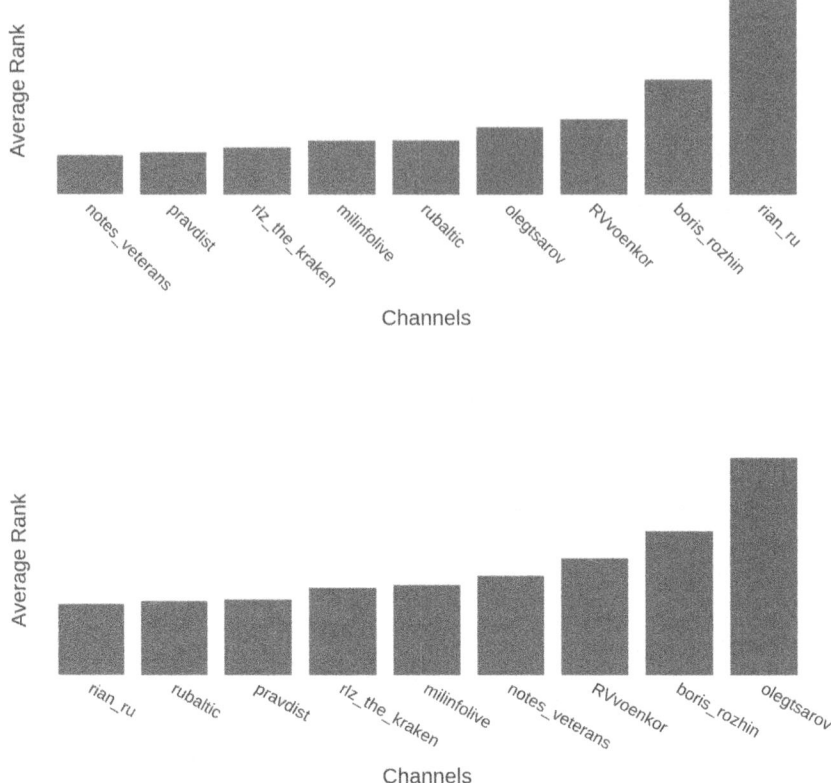

Fig. 4. The average ranks of disinformation detection across channels using SVO triples (top) and ERE triples (bottom). A lower number indicates a higher rank and a greater level of disinformation found by the metrics. The numbers on X axis are hidden because they do not provide any additional insight - they're simply averages of continuous ranks.

"riz_the_kraken" as having greater disinformation, whereas "boris_rozhin" and "RVvoenkor" were regarded as having lower rankings.

5.2 Detected Disinformation Cases

To detect disinformation, our five similarity metrics were normalized to the interval [0,1]. Scores below certain thresholds (0.8 – for the Maximum Common Subgraph (MCS), 0.5 – for the Subgraph Matching (SG), Weisfeiler Lehman (WL), and Neighborhood Hash (NH) Kernels, and 0.3 – for the Shortest Path (SP) Kernel) were excluded to focus on high similarity and reduce near-zero values.

Tables 1 and 2 show the 5 most discovered disinformation cases, which were then further explored via temporal analysis.

SVO and ERE triples reveal how these triple extraction approaches differ in representing text. Dependency parsing highlights the text's explicit linguistic

Table 1. Top 5 disinformation cases found via SVO triples

Disinformation cases	Occurrences
Recent strikes were aimed legitimate targets under the laws of war	114
Kherson, Zaporizhzhya, and Donbas regions overwhelmingly voted to join Russia	114
2 Ukrainian firms ordered to make dirty bomb, work near completion	114
The West is helping Ukraine create a "dirty bomb"	114
Kyiv and the West shot down flight MH17 in attempt to kill Putin or discredit Russia	114

structures and syntactic connections by capturing SVO triples. The claims concerning MH17, dirty bombs, and territorial referendums are among the direct and frequently sensational claims that are identified by this method because it tends to give priority to messages that closely match surface-level grammatical constructs [36] (see Table 1). Their grammatical prominence and repetition make these cases particularly noticeable.

Table 2. Top 5 disinformation cases found via ERE triples

Disinformation cases	Occurrences
The Kherson region, along with neighboring Zaporizhzhya and the Donetsk and Luhansk People's Republics, overwhelmingly voted to join Russia during a referendum in late September	82
Two Ukrainian firms have been ordered to produce a dirty bomb and the work is nearing completion	82
In the eight months since the beginning of the special military operation in Ukraine, the Russian Army restricted the operation to Ukrainian military targets, and avoided fighting the conflict in the same way that the Ukrainian military has since it began its terror bombing and shelling civilians in the Donbass in 2014	80
The West is increasing arms supplies and providing intelligence to Kiev's neo-Nazi regime	80
According to the available data, Ukraine is carrying out work to create a 'dirty' nuclear charge	80

ERE triple extraction, which focuses on semantic relations [37], captured conceptual connections and highlighted disinformation that included implicit or broader relational contexts overlooked by dependency parsing. Syntactically direct but contextually isolated claims were less prioritized than those expressing geopolitical claims or abstract associations (e.g., "Russia does not allow Kyiv to organize Holodomor in Ukraine," see Table 2).

These results highlight how different relational triple extraction methods shape KG creation and analysis. Dependency parsing often identifies explicit, repeated disinformation narratives, while ERE extraction can capture complex or more abstract patterns. Integrating both methods may provide a more comprehensive view of disinformation and detect a broader range of narratives.

5.3 Temporal Spread of Disinformation

To explore detected disinformation further, a temporal analysis was conducted to compare the results of two triple extraction techniques across all five graph distances to assess their similarity (Fig. 2). The timelines initially appeared quite different, e.g., sometimes moving in opposite directions, as with the MCS distance. However, closer examination of the distinct components (trend and seasonality; refer to Fig. 3) revealed that they were generally very similar. Except for MCS, which progressed in opposing directions, the seasonalities and trends for the SG, WL, and NH Kernels were almost the same across both triple extraction methods.

Comparing five graph similarity techniques showed that SP had the most peaks and detected the highest number of disinformation cases, making it the most reactive. SM and WL also identified a notable volume of disinformation, while MCS and NH were more stable and found fewer instances. MCS exhibited the largest discrepancy from the other metrics and was least effective. Consequently, SP was considered best for this task, with SM offering robust performance.

A separate temporal analysis identified peak disinformation periods across triple extraction approaches. Therefore, time frames matching the spikes across both of them were selected (Fig. 2), and the most frequent disinformation types were determined. All five graph similarity measures helped to pinpoint the main storylines, summarized below for each period:

- *2024 02 26 – 2025 03 04*. During this period, references to Flight MH17's downing and claims that the West had ordered dirty bombs were the most frequent disinformation narratives.
- *2024 05 22 – 2024 06 05*. During this period, "dirty bomb" and MH17 claims persisted, while discussions about Donbas, Zporizhzhya, and Kherson allegedly joining Russia surged.
- *2024 08 21 – 2024 08 28*. During this period, disinformation claimed Russia prepared its own weapons after Zelensky urged Western nuclear strikes and that the West supported a neo-Nazi government.
- *2024 11 11 – 2024 11 25*. During the final spike, dominant disinformation focused on MH17, claims that occupied regions wanted to join Russia, the idea that Ukraine was a Russian creation, and allegations that Ukrainian soldiers tried to seize the Zaporizhzhia power plant using human shields.

6 Conclusions

This study examined disinformation detection using two graph construction methods—syntactic dependency- and semantic relation-based—and five graph-based similarity measures. Over 100,000 messages from nine Russian Telegram channels were analysed for disinformation prevalence, detected cases, and temporal patterns.

Both graph construction methods produced similar yet distinct insights, influencing channel rankings regarding the volume of disinformation: "rian_ru", for instance, was flagged most disinformation-heavy by the syntactic method but least by the semantic one as the former prioritized surface-level grammatical constructs, while the latter favoured implicit or broader relational contexts overlooked by dependency parsing. Most channels, however, showed stable rankings across both triple extraction methods.

Among our five graph-based similarity measures, SP proved to be the most sensitive to varied events, detecting the highest disinformation volume, while SM showed consistency suitable for long-term analysis. Notable narratives emerged, e.g. Flight MH17, claims of Western assistance, and wishes of occupied regions to join Russia, highlighting how graph-based methods can reveal disinformation patterns in real-time communication.

Beyond these immediate results, the scalability of our method is a key advantage for large-scale analyses. Because knowledge graphs convert unstructured text into structured triples, we can rapidly expand the system to accommodate additional disinformation cases without excessive computational cost. Currently, the EUvsDisinfo database contains over 18,000 cases, and its daily updates allow seamless integration of new narratives. This real-time updating mechanism ensures that our approach can quickly capture and flag newly identified narratives as they emerge, thereby supporting continuous monitoring.

We chose this knowledge graph approach over more conventional machine learning or text-matching methods due to its ability to handle vast volumes of data with minimal resource expenditure. Traditional approaches often demand extensive labelled training sets and computationally expensive model updates. By contrast, our graph-based system primarily retains the structural elements of textual data, thus simplifying storage and accelerating disinformation detection. Nonetheless, it is important to note that this method is designed predominantly to reveal known disinformation narratives rather than to discover entirely novel ones.

Going forward, we envision further enhancements, such as integrating advanced knowledge graph construction techniques (including those powered by large language models) to capture deeper contextual cues and enrich the graph with additional metadata (e.g., propaganda or framing techniques). Such expansions could improve our ability to detect disinformation patterns that are currently missed by more narrowly focused syntactic or semantic approaches.

References

1. Arjunan, T.: Detecting anomalies and intrusions in unstructured cybersecurity data using natural language processing. Int. J. Res. Appl. Sci. Eng. Technol. (2024). https://doi.org/10.22214/ijraset.2024.58497
2. Artene, C.G., Oprisa, C., Buțincu, C., Leon, F.: Finding patient zero and tracking narrative changes in the context of online disinformation using semantic similarity analysis. Mathematics (2023). https://doi.org/10.3390/math11092053
3. Asghar, M.Z., Habib, A., Habib, A., Khan, A., Ali, R., Khattak, A.: Exploring deep neural networks for rumor detection. J. Ambient. Intell. Humaniz. Comput. **12**, 4315–4333 (2021)
4. Besta, M., et al.: Graph of thoughts: solving elaborate problems with large language models (2024). https://doi.org/10.48550/arXiv.2308.09687, http://arxiv.org/abs/2308.09687
5. Borgwardt, K., Ghisu, E., Llinares-López, F., O'Bray, L., Rieck, B., et al.: Graph kernels: state-of-the-art and future challenges. Found. Trends® Mach. Learn. **13**(5-6), 531–712 (2020)
6. Bytyci, A., Ramosaj, L., Bytyci, E.: Review of automatic and semi-automatic creation of knowledge graphs from structured and unstructured data. In: RTA-CSIT, pp. 72–79 (2023)
7. Cabot, P.L.H., Navigli, R.: REBEL: relation extraction by end-to-end language generation. In: Findings of the Association for Computational Linguistics: EMNLP 2021, pp. 2370–2381 (2021)
8. Chen, K.C., Chen, C.Y., Li, C.T.: Anti-disinformation: an adversarial attack and defense network towards improved robustness for disinformation detection on social media. In: 2023 IEEE International Conference on Big Data (BigData), pp. 5476–5484 (2023). https://api.semanticscholar.org/CorpusID:267149224
9. Erb, W., et al.: Krylov subspace methods to accelerate kernel machines on graphs. Adv. Comput. Sci. Eng. **1**(1), 59–81 (2023)
10. Gao, T., Yang, J., Peng, W., Jiang, L., Sun, Y., Li, F.: A content-based method for sybil detection in online social networks via deep learning. IEEE Access **8**, 38753–38766 (2020)
11. Gong, S., Sinnott, R.O., Qi, J., Paris, C.: Fake news detection through graph-based neural networks: a survey (2023). https://doi.org/10.48550/arXiv.2307.12639, http://arxiv.org/abs/2307.12639
12. Gupta, P., Nigam, S., Singh, R.: Automatic extractive text summarization using multiple linguistic features. ACM Trans. Asian Low-Resour. Lang. Inf. Process. (2024). https://doi.org/10.1145/3656471
13. Hannah, G., Payne, T.R., Mitchell, A., Piercy, E., Konev, B.: Towards a methodology for the semi-automatic generation of scientific knowledge graphs from xml documents. In: In OM@ ISWC, pp. 85–90 (2023)
14. Haussler, D.: Convolution kernels on discrete structures. Technical report UCSC-CRL-99-10, University of California, Santa Cruz (1999). http://www0.cs.ucl.ac.uk/staff/m.pontil/reading/haussler.pdf. Accessed 02 May 2024
15. He, X., Bresson, X., Laurent, T., Perold, A., LeCun, Y., Hooi, B.: Harnessing explanations: LLM-to-LM interpreter for enhanced text-attributed graph representation learning (2023). https://doi.org/10.48550/arXiv.2305.19523, http://arxiv.org/abs/2305.19523
16. Ibrahim, F., Aoun, M.: Improving query efficiency in heterogeneous big data environments through advanced query processing techniques. J. Contemp. Healthc. Anal. **6**(6), 40–64 (2022)

17. Jelodar, H., et al.: Latent Dirichlet allocation (LDA) and topic modeling: models, applications, a survey. Multimed. Tools Appl. **78**, 15169–15211 (2019)
18. Kaliyar, R.K., Goswami, A., Narang, P.: FakeBERT: fake news detection in social media with a BERT-based deep learning approach. Multimed. Tools Appl. **80**(8), 11765–11788 (2021)
19. Kriege, N.M., Johansson, F.D., Morris, C.: A survey on graph kernels **5**(1), 1–42 (2020). https://doi.org/10.1007/s41109-019-0195-3
20. Li, X., Fan, J.: Entity relationship extraction method based on dependency syntax analysis and rules. In: Proceedings of the 2019 International Conference on Robotics Systems and Vehicle Technology - RSVT 2019 (2019). https://doi.org/10.1145/3366715.3366740
21. Mishra, S., Shukla, P.K., Agarwal, R.: Location wise opinion mining of real time twitter data using hadoop to reduce cyber crimes. In: 2nd International Conference on Data, Engineering and Applications (IDEA), pp. 1–6. IEEE (2020). https://ieeexplore.ieee.org/document/9170700
22. Nascita, A., Montieri, A., Aceto, G., Ciuonzo, D., Persico, V., Pescapé, A.: XAI meets mobile traffic classification: Understanding and improving multimodal deep learning architectures. IEEE Trans. Netw. Serv. Manage. **18**(4), 4225–4246 (2021). https://ieeexplore.ieee.org/document/9490313
23. Nastase, V., Kotnis, B.: Abstract graphs and abstract paths for knowledge graph completion. In: Proceedings of the Eighth Joint Conference on Lexical and Computational Semantics (*SEM 2019), pp. 147–157. Association for Computational Linguistics (2019). https://doi.org/10.18653/v1/S19-1016, https://www.aclweb.org/anthology/S19-1016
24. Nikolentzos, G., Siglidis, G., Vazirgiannis, M.: Graph kernels: a survey. J. Artif. Intell. Res. **72**, 943–1027 (2021)
25. Opdahl, A.L., Al-Moslmi, T., Dang-Nguyen, D.T., Gallofré Ocaña, M., Tessem, B., Veres, C.: Semantic knowledge graphs for the news: a review. **55**(7), 140:1–140:38 (2022). https://doi.org/10.1145/3543508, https://dl.acm.org/doi/10.1145/3543508
26. Perez, R.C., Da Veiga, S., Garnier, J., Staber, B.: Gaussian process regression with sliced Wasserstein Weisfeiler-Lehman graph kernels. In: International Conference on Artificial Intelligence and Statistics, pp. 1297–1305. PMLR (2024)
27. Qin, Y., et al.: Entity relation extraction based on entity indicators. Symmetry **13**, 539 (2021). https://doi.org/10.3390/SYM13040539
28. Qin, Y., Dominik, W., Tang, C.: Predicting future rumours. Chin. J. Electron. **27**(3), 514–520 (2018). https://cje.ejournal.org.cn/en/article/doi/10.1049/cje.2018.03.008
29. Quer, S., Marcelli, A., Squillero, G.: The maximum common subgraph problem: a parallel and multi-engine approach. Computation **8**(2), 48 (2020)
30. Ranade, P., Dey, S., Joshi, A., Finin, T.: Computational understanding of narratives: a survey. IEEE Access **10**, 101575–101594 (2022). https://doi.org/10.1109/ACCESS.2022.3205314, https://ieeexplore.ieee.org/document/9882117
31. Salamanos, N., Leonidou, P., Laoutaris, N., Sirivianos, M., Aspri, M., Paraschiv, M.: HyperGraphDis: leveraging hypergraphs for contextual and social-based disinformation detection. In: Proceedings of the International AAAI Conference on Web and Social Media, vol. 18, pp. 1381–1394 (2024)
32. Saleem, S., Mehrotra, M.: Context-aware transfer learning approach to detect informative social media content for disaster management. Int. J. Adv. Comput. Sci. Appl. (2024). https://doi.org/10.14569/ijacsa.2024.0150167

33. Schulz, T.H., Horváth, T., Welke, P., Wrobel, S.: A generalized Weisfeiler-Lehman graph kernel. Mach. Learn. **111**(7), 2601–2629 (2022)
34. Shrivastava, G., Kumar, P., Ojha, R.P., Srivastava, P.K., Mohan, S., Srivastava, G.: Defensive modeling of fake news through online social networks. IEEE Trans. Comput. Soc.l Syst. **7**(5), 1159–1167 (2020)
35. Shu, K., Sliva, A., Wang, S., Tang, J., Liu, H.: Fake news detection on social media: a data mining perspective. ACM SIGKDD Explor. Newsl **19**(1), 22–36 (2017)
36. Takko, T., Bhattacharya, K., Lehto, M., Jalasvirta, P., Cederberg, A., Kaski, K.: Knowledge mining of unstructured information: application to cyber domain. Sci. Rep. **13**(1), 1714 (2023)
37. Tang, W., et al.: UniRel: Unified representation and interaction for joint relational triple extraction. In: Proceedings of the 2022 Conference on Empirical Methods in Natural Language Processing, pp. 7087–7099 (2022)
38. Tuo, M., Yang, W.: Review of entity relation extraction. J. Intell. Fuzzy Syst. **44**, 7391–7405 (2023). https://doi.org/10.3233/jifs-223915
39. Wu, J., Orlandi, F., O'Sullivan, D., Dev, S.: On the use of virtual knowledge graphs to improve environmental sensor data accessibility. IEEE J. Sel. Top. Appl. Earth Observ. Remote Sens. **17**, 6671–6682 (2024). https://doi.org/10.1109/JSTARS.2024.3370389
40. Zhou, X., Zafarani, R.: Fake news: a survey of research, detection methods, and opportunities. arXiv:1812.00315 [cs] **2** (2018). https://arxiv.org/abs/1812.00315

Computation of the Laplacian Spectral Barycentre Network in a Soules Basis

François G. Meyer

Applied Mathematics, University of Colorado at Boulder, Boulder, CO 80305, USA
fmeyer@colorado.edu
https://francoismeyer.github.io

Abstract. The main contribution of this work is a fast algorithm to compute the barycentre of a set of networks based on a Laplacian spectral pseudo-distance. The core engine for the estimation of the barycentre is an algorithm that explores the large library of Soules bases, and returns *the best Soules basis* that leads to the reconstruction of a weighted network whose spectrum is the sample mean spectrum, and whose geometry matches that of the sample mean adjacency matrix. We prove that when the networks are random realizations of stochastic block models our algorithm reconstructs the population mean adjacency matrix. In addition to the theoretical analysis, we perform Monte Carlo simulations to validate the theory. This work is significant because it opens the door to the design of new spectral-based network synthesis that have theoretical guarantees.

Keywords: Barycentre network · Soules basis · sample Fréchet mean

1 Introduction, Problem Statement, and Related Work

1.1 The Barycentre Network

The design of machine learning algorithms that can analyze "network-valued" random variables is of fundamental importance. Such machine learning algorithms often require the computation of a "sample mean" network that can summarize the topology and connectivity of a dataset of networks, $\{G^{(1)}, \ldots, G^{(N)}\}$. Formally, we denote by $\mathcal{S}$ the set of $n \times n$ symmetric adjacency matrices with nonnegative weights, and we assume that the adjacency matrix $\boldsymbol{A}^{(k)}$ of the network $G^{(k)}$ is sampled from a probability space $(\mathcal{S}, \mathbb{P})$. An example of a probability space is the stochastic block model (see Sect. 1.3). We equip the probability space $(\mathcal{S}, \mathbb{P})$ with a metric d to quantify proximity of networks. A notion of summary network is provided by the concept of *barycentre* network, $\widehat{\boldsymbol{\mu}}_N[\mathbb{P}]$, also called the sample Fréchet mean [16], which minimizes the sum of the squared distances to all the networks in the sample,

$$\widehat{\boldsymbol{\mu}}_N[\mathbb{P}] \stackrel{\text{def}}{=} \operatorname*{argmin}_{\boldsymbol{A} \in \mathcal{S}} \sum_{k=1}^{N} d^2(\boldsymbol{A}, \boldsymbol{A}^{(k)}). \tag{1}$$

In this work, we propose a fast algorithm to compute the barycentre of a set of networks based on a Laplacian spectral pseudo-distance.

Before continuing, we introduce some notations. We denote by $[n] \stackrel{\text{def}}{=} \{1,\ldots,n\}$. We define $\mathbf{1} \stackrel{\text{def}}{=} [1\cdots 1]^T$, and $\mathbf{J} = \mathbf{1}\mathbf{1}^T$. We use $\mathbf{A}$ to denote the adjacency matrix of a network G, and $\mathbf{D}$ to denote the diagonal degree matrix. The symmetric normalized adjacency matrix, $\mathbf{B} = (b_{ij})$, is defined by

$$b_{ij} \stackrel{\text{def}}{=} a_{ij}/\sqrt{d_i d_j} \text{ if } d_i d_j \neq 0; \text{ and } b_{ij} \stackrel{\text{def}}{=} 0 \text{ otherwise.} \quad (2)$$

The normalized Laplacian is defined by $\mathcal{L} \stackrel{\text{def}}{=} \text{Id} - \mathbf{B}$. We denote by $\boldsymbol{\lambda}(\mathcal{L}) = [\lambda_1, \ldots, \lambda_n]$ the *ascending* sequence of eigenvalues of $\mathcal{L}$.

1.2 The Laplacian Spectral Pseudo-Distance

The metric d in (1) influences the topological characteristics that $\widehat{\boldsymbol{\mu}}_N[\mu]$ inherits from $\{G^{(1)}, \ldots, G^{(N)}\}$ [16]. We advocate that the distance between networks should be evaluated in the spectral domain, by comparing the eigenvalues of the normalized Laplacian, $\mathcal{L}^{(k)}$, of the respective networks $G^{(k)}$. We define the Laplacian spectral pseudo-metric as

$$d(\mathcal{L}, \mathcal{L}') \stackrel{\text{def}}{=} \|\boldsymbol{\lambda}(\mathcal{L}) - \boldsymbol{\lambda}(\mathcal{L}')\|_2. \quad (3)$$

This pseudo-distance captures at multiple scales the structural and connectivity information in the networks [25]. Defining a pseudo-distance in the spectral domain alleviates the difficulty of solving the node correspondence problem, and in the case of the normalized Laplacian, it makes it possible to compare networks of different sizes, since $\lambda_i(\mathcal{L}), \lambda_i(\mathcal{L}') \in [0, 2]$. When the networks are realizations of a stochastic block model, the eigenvalues of $\mathcal{L}$ associated with each community are better separated from the bulk than the corresponding eigenvalues of the Laplacian $\mathbf{L} \stackrel{\text{def}}{=} \mathbf{D} - \mathbf{A}$ [6].

In spite of the advantages of the pseudo-metric d, the computation of (1) leads to two technical obstacles. The first challenge stems from the fact that d is defined in the spectral domain, but the optimization (1) takes place in $\mathcal{S}$. This leads to the definition of a *realizable* spectral sequence; we say that $\boldsymbol{\lambda}$ is realizable if there exists $\mathbf{A} \in \mathcal{S}$ whose Laplacian, $\mathcal{L}(\mathbf{A})$, satisfies $\boldsymbol{\lambda}(\mathcal{L}(\mathbf{A})) = \boldsymbol{\lambda}$. Further, we define $\mathcal{R}$ to be the *set of realizable sequences*. We can then formally rephrase the optimisation problem associated with the estimation of $\widehat{\boldsymbol{\mu}}_N[\mathbb{P}]$ in (1),

$$\boldsymbol{\lambda}(\widehat{\boldsymbol{\mu}}_N[\mathbb{P}]) = \underset{\boldsymbol{\lambda} \in \mathcal{R}}{\text{argmin}} \sum_{k=1}^{N} \|\boldsymbol{\lambda} - \boldsymbol{\lambda}(\mathcal{L}^{(k)})\|_2^2. \quad (4)$$

If we relax this minimization problem ($\boldsymbol{\lambda} \in \mathbb{R}^n$), then the solution to (4) is the sample mean $\widehat{\mathbb{E}}_N[\boldsymbol{\lambda}] \stackrel{\text{def}}{=} N^{-1}\sum_{k=1}^{N} \boldsymbol{\lambda}(\mathcal{L}^{(k)})$, which may be realizable if one can find an orthonormal basis $\boldsymbol{\Psi}$ such that $\boldsymbol{\Psi}\widehat{\mathbb{E}}_N[\boldsymbol{\lambda}]\boldsymbol{\Psi}^T \in \mathcal{S}$.

Which brings us to the second difficulty: the knowledge of the eigenvalues of the barycentre network, $\boldsymbol{\lambda}(\widehat{\boldsymbol{\mu}}_N[\mathbb{P}])$, is insufficient to reconstruct a network;

we need a set of eigenvectors that summarizes the distribution of eigenvectors associated with the respective $\mathcal{L}^{(q)}$. To address this problem, one can align the eigenvectors of the $A^{(q)}$ [11] (or $\mathcal{L}^{(q)}$ [24]). Others [10] have proposed numerical methods to find the best SBM (p, q, n) whose eigenvalues match $\widehat{\mathbb{E}}_N[\lambda]$. More generally, we are interested in the question of solving a symmetric nonnegative inverse eigenvalue problem [23].

1.3 The Stochastic Block Model

To provide theoretical guarantees for the algorithms presented in this paper, we analyse the algorithms when the networks are sampled from a stochastic block model (e.g. [1], and references therein). Stochastic block models provide universal approximants to networks and can be used as building blocks to analyse more complex networks (e.g. [2,10,18], and references therein).

We define the general stochastic block model SBM (p, q, n). Let $\{B_k\}, 1 \leq k \leq M$ be a partition of the vertex set $[n]$ into M blocks (or communities). We define the vector $p = [p_1, \cdots, p_M]$ to be the edge probabilities within each block, and q to be the edge probability between blocks. The entries $a_{ij} = a_{ji}, i < j$ of the adjacency matrix A are independent (up to symmetry) and are distributed with Bernoulli distributions with parameter p_m if i and j are in the same block B_m, and parameter q if i and j are in distinct blocks. We set $a_{ii} = 0$ (no self-loop).

We represent SBM (p, q, n) by the matrix of edge probabilities, or matrix of connection probabilities, $P \stackrel{\text{def}}{=} \mathbb{E}[A]$. We sometimes consider a balanced version of the model where all blocks have the same size, $|B_m| = n/M$, (in that case we assume without loss of generality that n is a multiple of M), and all the edge probabilities are equal, $p_1 = \cdots = p_M$.

1.4 Content of the Paper: Our Main Contributions

The main contribution of this work is a fast algorithm to compute the barycentre of a set of networks based on the spectral pseudo-distance (3). The core engine is an algorithm that explores the large library of Soules bases [9,13] (which is organized as a binary tree [8]), and returns a Soules basis that can be used to construct the normalized Laplacian of a network, whose eigenvalues are equal to the population mean spectrum. We prove that if the networks are random realizations of SBM (p, q, n), the algorithm reconstructs the population mean adjacency matrix associated with the SBM. This work is significant because it opens the door to the design of new spectral-based network synthesis [3,21] that have theoretical guarantees. We share our code to facilitate future work [14].

2 Soules Bases: Definition and Properties

Soules bases [8,13] were invented to provide a solution to the following symmetric nonnegative inverse eigenvalue problem: find an orthogonal matrix Ψ such that

$\boldsymbol{\Psi}$ diag $(\lambda_1, \ldots, \lambda_n) \boldsymbol{\Psi}^T \in \mathcal{S}$. Soules bases provide a large family of solutions to this problem.

2.1 Definition of Soules Bases

A Soules bases is an orthogonal matrix that is constructed iteratively by applying a product of Givens rotations to a fixed vector $\boldsymbol{\psi}_1$ with nonnegative entries. The construction starts at the coarsest level ($l = 1$) with a normalized vector $\boldsymbol{\psi}_1$ with nonnegative entries, whose support is the interval $I^{(1)} = [n]$. Hereupon, our analysis assumes that we always choose $\boldsymbol{\psi}_1 \stackrel{\text{def}}{=} n^{-1/2} \mathbf{1}$.

At any given level l, the set $[n]$ is partitioned into l ordered intervals $I_q^{(l)}, 1 \leq q \leq l$. When progressing from level l to $l + 1$, one chooses an interval, $I_q^{(l)} = [i_0, i_1)$, and one chooses an index $k \in [i_0, i_1]$ and defines $I_q^{(l+1)} \stackrel{\text{def}}{=} [i_0, k]$, and $I_{q+1}^{(l+1)} \stackrel{\text{def}}{=} [k + 1, i_1]$ (see Fig. 1-right). The split of $I_q^{(l)}$ into $I_q^{(l+1)}$ and $I_{q+1}^{(l+1)}$ triggers the construction of the Soules vector $\boldsymbol{\psi}_{l+1}$ (see Fig. 1-left), defined by

$$\psi_{l+1}(i) \stackrel{\text{def}}{=} \frac{1}{\sqrt{\|\psi(i_0 : i_1)\|}} \begin{cases} \frac{\|\psi_1(k+1:i_1)\|}{\|\psi_1(i_0:k)\|} \psi_1(i) & \text{if } i_0 \leq i \leq k \\ -\frac{\|\psi_1(i_0:k)\|}{\|\psi_1(k+1:i_1)\|} \psi_1(i) & \text{if } k+1 \leq i \leq i_1, \\ 0 & \text{otherwise} \end{cases} \quad (5)$$

where the vectors $\boldsymbol{\psi}_1(i_0 : i_1), \boldsymbol{\psi}_1(i_0 : k)$, and $\boldsymbol{\psi}_1(k + 1 : i_1)$ are n-dimensional vectors whose nonzero entries are extracted from $\boldsymbol{\psi}_1$ at the corresponding indices,

$$\begin{aligned} \boldsymbol{\psi}_1(i_0 : i_1) &= [0 \cdots 0 \ \psi_1(i_0) \cdots \psi_1(k) \ \psi_1(k+1) \cdots \psi_1(i_1) \ 0 \cdots 0]^T, \\ \boldsymbol{\psi}_1(i_0 : k) &= [0 \cdots 0 \ \psi_1(i_0) \cdots \psi_1(k) \ 0 \cdots\cdots\cdots\cdots\cdots\cdots\cdots 0]^T, \quad (6) \\ \boldsymbol{\psi}_1(k+1 : i_1) &= [0 \cdots\cdots\cdots\cdots\cdots\cdots\cdots 0 \ \psi_1(k+1) \cdots \psi_1(i_1) \ 0 \cdots 0]^T. \end{aligned}$$

The iterative subdivision process can be described using a binary tree (see Fig. 2-right) where a new vector is created at each node that has two children. We observe that $\boldsymbol{\psi}_l$ and $\boldsymbol{\psi}_m$, $l \neq m$, are either nested, or they do not overlap; whence $\langle \boldsymbol{\psi}_l, \boldsymbol{\psi}_m \rangle = 0$, and $[\boldsymbol{\psi}_1 \cdots \boldsymbol{\psi}_n]$ is an orthonormal matrix [8]; see [22] for a similar construction of Walsh-Hadamard packets.

2.2 Properties of Soules Bases

Using (5), we derive the following lemma with a proof by induction.

Lemma 1 (See [8]). *Let $[\boldsymbol{\psi}_1 \cdots \boldsymbol{\psi}_n]$ be a Soules basis constructed according to (5). Then,*

$$\forall m = 1, \ldots, n, \quad \boldsymbol{E}_m \stackrel{\text{def}}{=} \sum_{q=1}^{m} \boldsymbol{\psi}_q \boldsymbol{\psi}_q^T \geq 0, \text{ and } \boldsymbol{E}_n = \mathrm{Id}. \quad (7)$$

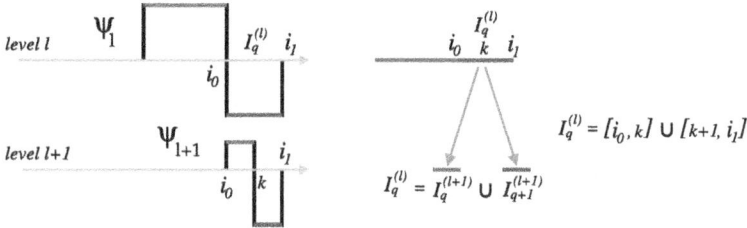

Fig. 1. Left: ψ_{l+1} is created by splitting the block of indices $I_q^{(l)} = [i_0, i_1]$ at level l into two sub-blocks, $[i_0, k] \cup [k+1, i_1]$ at level $l+1$. Right: a node in the Soules binary tree is triggered by the splitting of $[i_0, i_1] = [i_0, k] \cup [k+1, i_1]$.

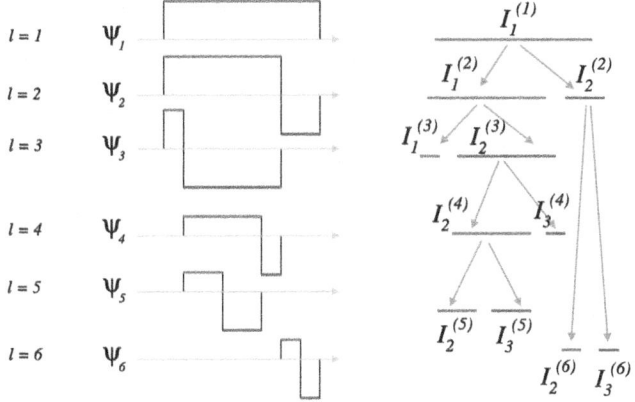

Fig. 2. Left: starting from level $l = 1$, one Soules vector ψ_l is constructed at each level $l \geq 2$ by selecting and then splitting an interval $I_q^{(l)}$ over which an already existing vector ψ_m, $m \leq l$ keeps a constant value. Right: each Soules basis is associated with a binary tree. The leaves of the tree are intervals that are not split.

Finally, we have the fundamental property of Soules bases.

Lemma 2 (See [8]). *Let $\boldsymbol{\Psi}$ be a Soules basis constructed according to (5). Let $\boldsymbol{\Lambda} = \mathrm{diag}(\lambda_1, \ldots, \lambda_n)$, where $\lambda_1 \geq \lambda_2 \geq \cdots \geq \lambda_n$. Then, the off-diagonal entries of $\boldsymbol{\Psi}\boldsymbol{\Lambda}\boldsymbol{\Psi}^T$ are non-negative. In addition, if $\lambda_n \geq 0$, then $\boldsymbol{\Psi}\boldsymbol{\Lambda}\boldsymbol{\Psi}^T \geq 0$.*

We note that there has been some recent interest in Soules bases to solve various inverse eigenvalue problems [7,19].

Remark 1. The result in Lemma 2 relies on the fact that the sequence of eigenvalues is decreasing. On the other hand, the eigenvalues of $\mathcal{L}$ are by nature ranked in ascending order (the index k of eigenvalue λ_k of $\mathcal{L}$ encodes the frequency of the corresponding eigenvector). Given an ascending sequence of eigenvalues of $\mathcal{L}$, $0 = \lambda_1 < \lambda_2 \leq \cdots \leq \lambda_n$, we would like to apply Lemma 2 to reconstruct a Laplacian matrix using a Soules basis. Since the off-diagonal entries of a normalized Laplacian $\mathcal{L}$ are nonpositive, we need to work with $-\boldsymbol{\lambda}$. Then,

$0 = \lambda_1 > -\lambda_2 \geq \cdots \geq -\lambda_n$, and we can use Lemma 2 to construct $\widehat{\mathcal{L}}$ such that

$$\widehat{\mathcal{L}} = \boldsymbol{\Psi} \operatorname{diag}(\lambda_1, \ldots, \lambda_n) \boldsymbol{\Psi}^T, \quad \text{where } \widehat{\mathcal{L}}_{ij} \leq 0 \text{ if } i \neq j. \tag{8}$$

Since we choose, $\boldsymbol{\psi}_1 = n^{-1/2} \mathbf{1}$, we have $\widehat{\mathcal{L}} \mathbf{1} = \mathbf{0}$, and therefore $\widehat{\mathcal{L}}_{ii} \geq 0$. While the signs of the entries of $\widehat{\mathcal{L}}$ match those of a normalized Laplacian, there is no guarantee that $\widehat{\mathcal{L}}$ be a valid normalized Laplacian (but see a definite answer in the case of the combinatorial Laplacian, $\boldsymbol{L} = \boldsymbol{D} - \boldsymbol{A}$ in [7]). This remark notwithstanding, we settle this question in Sect. 4, in the case where all the networks are sampled from SBM $(\boldsymbol{p}, q, n)$.

3 A Soules Basis for the Sparse Approximation of $\widehat{\mathbb{E}}_N[\boldsymbol{A}]$

3.1 Pre-processing Step: Approximation of the Networks

Approximation of Each Network with a Piecewise Constant Graphon. The first step of the computation of the barycentre graph involves the approximation of each of the N networks, $G^{(q)}$, $1 \leq q \leq N$, with step graphons (e.g., [2,4,10]). To facilitate the exposition, we assume that we can use a common SBM $(\boldsymbol{p}, q, n)$.

Spectral Clustering of the Nodes. Our approach does not assume that the graphon approximations are "well-aligned" – in the sense that nodes are clustered in blocks wherein the step graphon takes a constant value. We therefore need to cluster the nodes into blocks;. we use a spectral clustering method based on the eigenvectors of the normalized graph Laplacian [15,20] to organize nodes into clusters (the algorithm only requires the knowledge of the number of clusters). The inherent uncertainty associated with the cluster labels is resolved by ranking the clusters according to their volume.

3.2 Description of the Best Soules Basis Algorithm

After the pre-processing step, we assume that each $\boldsymbol{A}^{(k)}$ is the adjacency matrix of the SBM approximation of the network $G^{(k)}$; the vertices of each $G^{(k)}$ are rearranged in $\boldsymbol{A}^{(k)}$ according to the canonical ordering associated with the partition of $[n]$ into the blocks defined by SBM $(\boldsymbol{p}, q, n)$. To simplify the exposition, we assume hereupon that the sample of N graphs is composed of independent realizations of SBM $(\boldsymbol{p}, q, n)$, $G^{(1)}, \ldots, G^{(N)}$ of SBM $(\boldsymbol{p}, q, n)$. We define the sample edge probability matrix $\widehat{\mathbb{E}}_N[\boldsymbol{A}] \stackrel{\text{def}}{=} N^{-1} \sum_{q=1}^N \boldsymbol{A}^{(q)}$.

Given $\widehat{\mathbb{E}}_N[\boldsymbol{A}]$, the *best Soules algorithm* explores iteratively the binary tree of Soules vectors from the top level to the bottom level. The objective is to assemble a basis that provides a sparse approximation of $\widehat{\mathbb{E}}_N[\boldsymbol{A}]$. At each level l, the algorithm searches through all possible splits to find a new Soules vector

ψ_{l+1} that minimizes the residual approximation error between $\widehat{\mathbb{E}}_N[A]$ and its expansion in the first $l+1$ Soules vectors, $\psi_1, \ldots, \psi_{l+1}$,

$$\psi_{l+1} = \underset{\substack{\psi_{l+1} \\ \text{defined} \\ \text{by}(5)}}{\operatorname{argmin}} \left\| \widehat{\mathbb{E}}_N[A] - \sum_{q=1}^{l+1} \langle \psi_q \psi_q^T, \widehat{\mathbb{E}}_N[A] \rangle \psi_q \psi_q^T \right\|_F^2. \tag{9}$$

3.3 Theoretical Guarantees for the Best Soules Basis Algorithm

Our analysis of the best Soules basis is performed under the assumption that the input to the algorithm is not the sample mean adjacency matrix $\widehat{\mathbb{E}}_N[A]$ but its population equivalent $\mathbb{E}[A] = P$. This assumption is realistic if the network size n is large enough for some concentration phenomenon to be in effect. Our experiments (see Fig. 6-left) confirm the validity of this assumption. A finite sample analysis of the error bounds is left for future work.

The following lemma proves that the first M vectors of the best Soules basis recover the geometry of the SBM.

Lemma 3. *Let P be the population mean adjacency matrix of SBM (p, q, n) defined by*

$$P = \sum_{m=1}^{M} (p_m - q) \mathbf{1}_{B_m} \mathbf{1}_{B_m}^T + qJ, \tag{10}$$

where the M blocks B_m form a partition of $[n]$. Let $J_l, 1 \leq l \leq M$ be the leaves in the binary Soules tree (these are intervals that are no longer split, see Fig. 2-right) after M steps of the best Soules basis algorithm. Then, the blocks $\{B_m\}$ in (10) coincide with the intervals $\{J_l\}$. The entries of the matrix $E_M = \sum_{m=1}^{M} \psi_m \psi_m^T$ satisfy

$$e_M(i,j) = \begin{cases} \dfrac{1}{|J_m|} & \text{if } (i,j) \in J_m \times J_m, \\ 0 & \text{otherwise,} \end{cases} \tag{11}$$

where $|J_m|$ is the length of the interval J_m.

Proof. The full detailed rigorous proof of Lemma 3 is provided in the supplementary material [17]; we give here only the key ingredients of the proof. The first result concerns the matrix E_M.

Lemma 4 (See [17]). *Let J_m be the leaves in the binary Soules tree (these are intervals that are no longer split) after M steps of the best Soules basis algorithm. Then, E_M is given by*

$$e_M(i,j) = \begin{cases} \dfrac{1}{|J_l|} & \text{if } (i,j) \in J_l \times J_l, \\ 0 & \text{otherwise.} \end{cases} \tag{12}$$

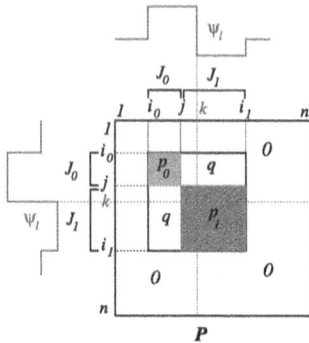

Fig. 3. The vector ψ_l (in blue) is created by splitting a block of indices $[i_0, i_1]$ at level $l-1$ into two sub-blocks, $[i_0, k] \cup [k+1, i_1]$ at level l. We consider the matrix P that is nonzero only on $[i_0, i_1] \times [i_0, i_1]$, and is piecewise constant on two blocks $J_0 \times J_0$ (in green) and $J_1 \times J_1$ (in red), where $J_0 = [i_0, j]$, and $J_1 = [j+1, i_1]$ (Color figure online)

We can derive Lemma 4 with a proof by induction on M, using the construction of ψ_l in (5).

The second ingredient of the proof of Lemma 3 specifically addresses the construction of each vector ψ_m in the best Soules basis. We can prove that at each level l, the Soules vector ψ_l is aligned with the boundary of a block B_m of the edge probability matrix P defined by (10). The proof hinges on the study of one iteration of the best Soules basis algorithm, as explained in Lemma 5. The proof of Lemma 5 is a simple calculation that relies on the fact that both $\psi_l \psi_l^T$ and P are piecewise constant over $[i_0, i_1] \times [i_0, i_1]$ (see Fig. 3).

Lemma 5 (See [17]). *Let ψ_l be the best Soules vector at level l with support* $\mathrm{supp}(\psi_l) \stackrel{def}{=} [i_0, i_1]$. *We consider the matrix P that is nonzero only on $[i_0, i_1] \times [i_0, i_1]$, and is piecewise constant on the two blocks $J_0 \times J_0$ and $J_1 \times J_1$, where $J_0 = [i_0, j]$, and $J_1 = [j+1, i_1]$ (see Fig. 3),*

$$P = p_0(\mathbf{1}_{J_0}\mathbf{1}_{J_0}^T) + p_1(\mathbf{1}_{J_1}\mathbf{1}_{J_1}^T) + q(\mathbf{1}_{J_0}\mathbf{1}_{J_1}^T + \mathbf{1}_{J_1}\mathbf{1}_{J_0}^T). \tag{13}$$

Then, $|\langle \psi_l \psi_l^T, P \rangle|^2$ is maximum if the location of the zero-crossing of ψ_l is equal to the location of the jump in the SBM, $k = j$ (see Fig. 3).

Lemma 6 extends Lemma 5 to the edge probability matrix P of a general SBM; it can be proved using a proof by contradiction (using Lemma 5).

Lemma 6 (See [17]). *Let P be the population mean adjacency matrix of* $\mathrm{SBM}(\boldsymbol{p}, q, n)$ *defined by*

$$P = \sum_{m=1}^{M}(p_m - q)\mathbf{1}_{B_m}\mathbf{1}_{B_m}^T + qJ, \tag{14}$$

where the M blocks B_m form a partition of $[n]$. Then, the split that creates ψ_2, is always located at the boundary between two blocks B_m and B_{m+1}.

***M* Iterations of the Best Soules Basis Algorithm.** This last lemma guarantees that after M iterations of algorithm, the matrix $\boldsymbol{E}_M$ associated with the first M Soules vectors recovers the block geometry. Lemma 7 is proved by induction on M, using Lemma 6.

Lemma 7 (See [17]). *Let $\boldsymbol{P}$ be the population mean adjacency matrix of SBM $(\boldsymbol{p}, q, n)$ defined by*

$$\boldsymbol{P} = \sum_{m=1}^{M} (p_m - q) \mathbf{1}_{B_m} \mathbf{1}_{B_m}^T + q\boldsymbol{J}. \qquad (15)$$

Let $J_l, 1 \leq l \leq M$ be the leaves in the binary Soules tree (these are intervals that are no longer split) after M steps of the best Soules basis algorithm. Then, the M blocks $\{B_m\}$ in (15) coincide with the M intervals $\{J_l\}$ discovered by the algorithm.

Finally, Lemma 3 is a direct consequence of Lemma 7 and Lemma 4. □

4 The Reconstruction of the Normalized Laplacian

4.1 Description of the Algorithm

We assume that we have access to N independent realizations of SBM $(\boldsymbol{p}, q, n)$. For each realization $G^{(q)}$, we compute the eigenvalues, $\boldsymbol{\lambda}(\boldsymbol{\mathcal{L}}^{(q)})$, of the normalized Laplacian $\boldsymbol{\mathcal{L}}^{(q)}$. We estimate the sample mean spectrum of the normalized Laplacian,

$$\begin{bmatrix} \overline{\lambda}_1 \ldots \overline{\lambda}_n \end{bmatrix}^T \overset{\text{def}}{=} \widehat{\mathbb{E}}_N \left[\boldsymbol{\lambda}(\boldsymbol{\mathcal{L}}) \right] = \frac{1}{N} \sum_{q=1}^{N} \boldsymbol{\lambda}(\boldsymbol{\mathcal{L}}^{(q)}). \qquad (16)$$

In the following, we present an algorithm to construct a matrix $\widehat{\boldsymbol{\mathcal{L}}}$, whose spectrum is the sample mean spectrum $\widehat{\mathbb{E}}_N \left[\boldsymbol{\lambda}(\boldsymbol{\mathcal{L}}) \right]$. We prove that if the networks are sampled from a balanced SBM with $p_1 = \cdots = p_M$ and $\widehat{\mathbb{E}}_N \left[\boldsymbol{\lambda}(\boldsymbol{\mathcal{L}}) \right] \approx \mathbb{E}[\boldsymbol{\lambda}(\boldsymbol{\mathcal{L}})]$, then $\widehat{\boldsymbol{\mathcal{L}}} = \boldsymbol{\mathcal{L}}(\boldsymbol{P})$, the Laplacian of the population mean adjacency matrix. Our approach combines the structural information about geometry of the block in the SBM, which is provided by Lemma 3, along with the spectral information provided by the sample mean spectrum. Unfortunately, this solution, while theoretically satisfying, is numerically unstable. We propose a second estimator, which is numerically stable and has similar theoretical guarantees.

4.2 The Complete Reconstruction

As explained in Lemma 3, the best Soules basis $\boldsymbol{\Psi} = \begin{bmatrix} \boldsymbol{\psi}_1 \cdots \boldsymbol{\psi}_n \end{bmatrix}$ recovers the geometry of the blocks of SBM $(\boldsymbol{p}, q, n)$; we propose to use $\boldsymbol{\Psi}$ to design the following estimator of the normalized Laplacian,

$$\widehat{\boldsymbol{\mathcal{L}}} \overset{\text{def}}{=} \sum_{q=1}^{n} \overline{\lambda}_q \boldsymbol{\psi}_q \boldsymbol{\psi}_q^T. \qquad (17)$$

Theoretical Guarantees for The Reconstruction. Because the ψ_q are Soules vectors, the spectrum of $\widehat{\mathcal{L}}$ always coincides with $[\overline{\lambda}_1 \ldots \overline{\lambda}_n]^T$. We can prove more; when the networks are sampled from of a balanced SBM where $p_1 = \cdots = p_M$ and in the limit of large network size ($\widehat{\mathbb{E}}_N [\boldsymbol{\lambda}(\mathcal{L})] \approx \mathbb{E}[\boldsymbol{\lambda}(\mathcal{L})]$), then $\widehat{\mathcal{L}}$ is equal to $\mathcal{L}(\boldsymbol{P})$, the normalized Laplacian associated with the edge probability matrix $\boldsymbol{P}$.

Lemma 8. *Let $\boldsymbol{P}$ be the edge probability matrix of a balanced SBM $(\boldsymbol{p}, q, n)$ where $p_1 = \cdots = p_M = p$,*

$$\boldsymbol{P} = \sum_{m=1}^{M} (p-q) \mathbf{1}_{B_m} \mathbf{1}_{B_m}^T + q \boldsymbol{J}. \tag{18}$$

Then, the estimator $\widehat{\mathcal{L}}$ defined in (17) is given by

$$\widehat{\mathcal{L}} = \mathrm{Id} - \left\{ \frac{(p-q)}{p + (M-1)q} \left(\sum_{m=1}^{M} \boldsymbol{\psi}_m \boldsymbol{\psi}_m^T \right) + \frac{Mq}{p + (M-1)q} \boldsymbol{\psi}_1 \boldsymbol{\psi}_1^T \right\}, \tag{19}$$

and therefore $\widehat{\mathcal{L}} = \mathcal{L}(\boldsymbol{P})$.

The proof of Lemma 8, which is provided in the supplementary material [17], relies on estimates for the dominant eigenvalues of the symmetric normalized adjacency matrix $\boldsymbol{B}$ when $\boldsymbol{A}$ is the adjacency matrix of a network sampled from a balanced SBM $(\boldsymbol{p}, q, n)$ with M blocks [12].

4.3 A Partial Reconstruction

In practice, the estimator of the normalized Laplacian given by $\widehat{\mathcal{L}}$ in (17) is very poor. This numerical problem is perfectly natural: the geometry and edge density of the SBM is unfortunately encoded by the smallest eigenvalues of $\mathcal{L}$ (see Lemma 3). The full expansion provided by (17) is plagued by the largest eigenvalues of $\mathcal{L}$, which come from the bulk created by the stochastic nature of the model. This issue is exacerbated by the fact that the high frequency eigenvectors ($\boldsymbol{\psi}_l$ with large l) have small support and therefore are localized around fine scale random structures present in the sample mean adjacency matrix, $\widehat{\mathbb{E}}_N [\boldsymbol{A}]$. The expression (19) indicates that $\mathcal{L}(\boldsymbol{P})$ depends only on the first M eigenvectors of the Soules basis. Whence, we define the following estimator of the symmetric normalized adjacency matrix, $\boldsymbol{B}(\boldsymbol{P})$,

$$\widehat{\boldsymbol{B}} \stackrel{\mathrm{def}}{=} \frac{(p-q)}{p + (M-1)q} \left(\sum_{m=1}^{M} \boldsymbol{\psi}_m \boldsymbol{\psi}_m^T \right) + \frac{Mq}{p + (M-1)q} \boldsymbol{\psi}_1 \boldsymbol{\psi}_1^T, \tag{20}$$

so that $\mathcal{L} = \mathrm{Id} - \widehat{\boldsymbol{B}}$. As is shown in Lemma 8, $\widehat{\boldsymbol{B}} = \boldsymbol{B}(\boldsymbol{P})$ when the network is a balanced SBM $(\boldsymbol{p}, q, n)$ with $p_1 = \cdots = p_M = p$. For a general SBM, one therefore proposes to estimate $\mathcal{L}$ using a truncated reconstruction given by

$$\widehat{\mathcal{L}}_M \stackrel{\mathrm{def}}{=} \sum_{q=1}^{M} \overline{\lambda}_q \boldsymbol{\psi}_q \boldsymbol{\psi}_q^T. \tag{21}$$

In practice, one needs to estimate M, the number of eigenvalues outside the bulk. Fortunately, many estimators are available (e.g. [5, 26], and references therein).

Theoretical Guarantees for the Reconstruction. In the case of a balanced SBM $(\boldsymbol{p}, q, n)$ where $p_1 = \cdots = p_M = p$, a simple calculation reveals that

$$\widehat{\boldsymbol{B}} = \boldsymbol{E}_M - \widehat{\boldsymbol{\mathcal{L}}}_M. \tag{22}$$

We recall (see Lemma 3) that $\boldsymbol{E}_M$ is zero outside of the blocks $B_k \times B_k$ of the SBM. We can therefore interpret (22) as providing the reconstruction of $\boldsymbol{B}(\boldsymbol{P})$ given by (20) inside the blocks of the SBM, once the offset given by $\boldsymbol{E}_M$ has been removed.

4.4 The Reconstruction of the Adjacency Matrix

An Estimator of the Degree Matrix. In order to recover the expected adjacency matrix, $\boldsymbol{P}$, one needs an estimate, $\widehat{\boldsymbol{D}}$, of the degree matrix to compute $\widehat{\boldsymbol{D}}^{1/2}\widehat{\boldsymbol{B}}\widehat{\boldsymbol{D}}^{1/2}$. Lemma 3 yields an estimate of the location of the blocks in the SBM. Our estimate of the degree matrix, $\widehat{\boldsymbol{D}}$, is computed by averaging the degrees of the nodes in the sample mean adjacency matrix, $\widehat{\mathbb{E}}_N[\boldsymbol{A}]$, separately within each block,

$$\widehat{d}_i \stackrel{\text{def}}{=} \sum_{j \in J_m} \left[\widehat{\mathbb{E}}_N[\boldsymbol{A}]\right]_{ij} \quad \text{if } i \in J_m, \ 1 \leq m \leq M. \tag{23}$$

The estimator $\widehat{\boldsymbol{D}}$ proves to be extremely precise.

An Estimator of the Adjacency Matrix. We combine $\widehat{\boldsymbol{B}}$ given by (20) with $\widehat{\boldsymbol{D}}$ given by (23), to derive an estimator of the edge probability matrix,

$$\widehat{\boldsymbol{P}} \stackrel{\text{def}}{=} \widehat{\boldsymbol{D}}^{1/2}\widehat{\boldsymbol{B}}\widehat{\boldsymbol{D}}^{1/2}. \tag{24}$$

5 Experiments

We compare our theoretical analysis to finite sample estimates, which were computed using numerical simulations. The software used to conduct the experiments is publicly available [14]. All networks were generated using the SBM $(\boldsymbol{p}, q, n)$ model. The nodes of the random realizations of the adjacency matrices are permuted with a different random permutation for each realization (e.g., see Fig. 4-left). This random permutation only affects the estimation of the sample mean matrix (the sample mean spectrum is not affected).

We first illustrate the selection of the Soules vectors (guaranteed by Lemma 3). For this experiment, we use $M = 4$ communities of sizes $63, 147, 105, 197$ (see Fig. 4-right). The edge probabilities were given by $p_i = c_i \log n^2/n$, where the scaling factor c_i was chosen randomly in $[1, 4]$, and $q = 2 \log n/n$.

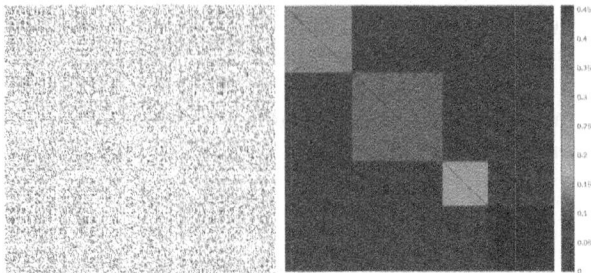

Fig. 4. Left: a random (unclustered) realization of the SBM (p, q, n) model. Right: original edge probability matrix P; We have $M = 4$ communities of sizes $67, 133, 71, 241$, the network size is $n = 512$; the edge probability within community i was $p_i \propto (\log n)^2/n$, The edge probability across communities was $q = 2(\log n)/n$.

We generate a randomly permuted single realisation of the SBM ($N = 1$) (see Fig. 4-left). This is clearly the least favorable scenario, where we expect that the estimation of the Soules basis is the most challenging. As shown in Fig. 5-left, the first three non trivial Soules vectors accurately detected the boundaries between the blocks, in spite of the very low contrast between the communities (see Fig. 4-left). This numerical evidence supports the theoretical analysis of Lemma 3. We then evaluated the accuracy of (24). Figure 4-right displays the original edge probability matrix P. Figure 5 displays the reconstructed adjacency matrix $\widehat{P}$ (center) using the top $M = 4$ Soules vectors, and the residual error $P - \widehat{P}$ (right). The mean absolute error, $n^{-2}\|P - \widehat{P}\|_1 \stackrel{\text{def}}{=} \frac{1}{n^2}\sum_{i=1}^{n}\sum_{j=1}^{n} |p_{ij} - \widehat{p}_{ij}|$, was $1.03e - 05$.

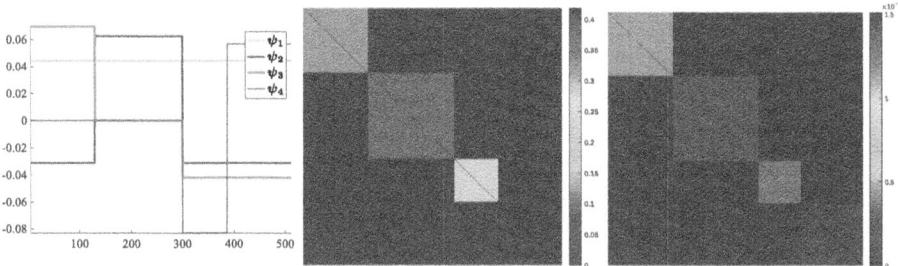

Fig. 5. Left: the first four trivial Soules vectors accurately detected the boundaries between the blocks, in spite of the very low contrast between the communities (see Fig. 4-left); center: the adjacency matrix of the barycentre network $\widehat{P}$, given by (24); right:the residual error between P and $\widehat{P}$. The mean absolute error was $n^{-2}\|P - \widehat{P}\|_1 = 1.03e - 05$.

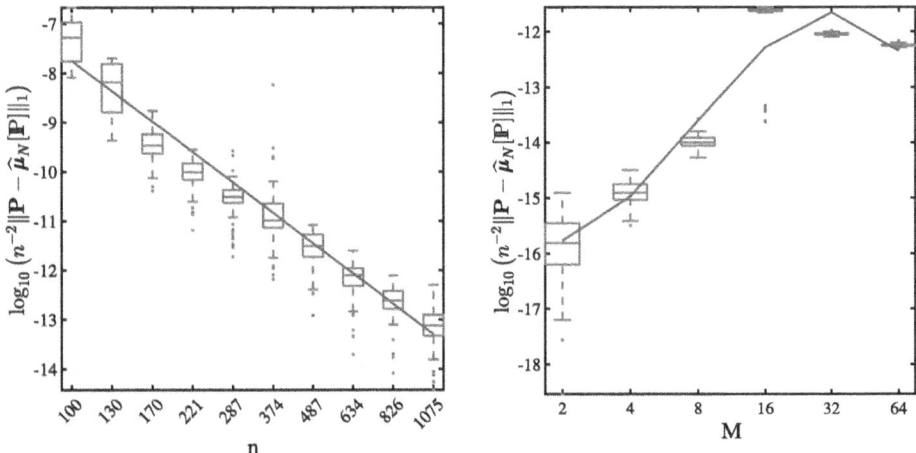

Fig. 6. Left: mean absolute error $n^{-2}\|\boldsymbol{P} - \widehat{\boldsymbol{P}}\|_1$ as a function of the network size, n. The network is composed of $M = 4$ communities, and is a scaled version of the network shown in Fig. 4-right. Right: mean absolute error $n^{-2}\|\boldsymbol{P} - \widehat{\boldsymbol{P}}\|_1$ as a function of the number of blocks, M. Each network is sampled from a balanced SBM $(\boldsymbol{p}, q, n)$ with M blocks of size n/M; $p_i = 3(\log n)^2/n$, $q = 2\log n/n$, and $n = 1{,}024$.

Next, we studied the effect of the network size, n, on the mean absolute reconstruction error (see Fig. 6-left). We rescaled the $M = 4$ SBM model described in the previous paragraph, keeping the relative sizes of the communities the same, and increased the network size from $n = 100$ to $n = 1{,}075$. For each n, we computed the mean absolute reconstruction error. As expected, the error decreases as a function of n. We found $n^{-2}\|\boldsymbol{P} - \widehat{\boldsymbol{P}}\|_1 \propto n^{-1.6}$. This experiment validates the theoretical derivations that were obtained in the limit of large network sizes, when some concentration phenomenon is in effect, and we can replace $\widehat{\mathbb{E}}_N[\boldsymbol{A}]$ with $\mathbb{E}[\boldsymbol{A}] = \boldsymbol{P}$ in our analysis of the best Soules basis algorithm.

We note that the alignment performed by the spectral clustering is not always accurate (as reflected in some outlier values in the mean absolute error; e.g., $n = 374$ in Fig. 6-left). This is due to the fact that we use a simple spectral clustering algorithm. The present work focuses on the construction of the barycentre network; the study of the combined performance of the pre-processing clustering step with the computation of the barycentre is left for future work. Fortunately, the computation of the best Soules basis only relies on the coarse scale eigenvectors. These eigenvectors are estimated by integrating the noisy estimate of the sample mean adjacency matrix, a process which effectively reduces the errors in the alignment.

The next experiment illustrates the effect of the number of blocks M in a balanced SBM $(\boldsymbol{p}, q, n)$ when the edge probabilities are equal, $p_1 = \cdots = p_M$. When M becomes large, then the first $M - 1$ non trivial eigenvalues λ_m of $\mathcal{L}$, converge to 1. Because these eigenvalues are no longer separated from the bulk, the truncated reconstruction (20) becomes numerically unstable, and the reconstruction error increases (see Fig. 6-right).

6 Discussion

We described a fast algorithm to compute the barycentre of a set of networks based on the Laplacian spectral pseudo-distance. An original contribution is an algorithm that explores the large library of Soules bases, and returns a basis that can be used to construct the normalized Laplacian of a network, whose eigenvalues are equal to the population mean spectrum. Our work is significant because not only do we match the eigenvalues, but we recover the community structure present in the network by means of the coarse scale Soules vectors. We are not aware of any work that takes advantage of the binary tree structure of the Soules bases; the present paper offers for the first time a principled method to quickly explore the large library of Soules bases.

Acknowledgement. The author is grateful to the anonymous reviewers for their insightful comments and suggestions that greatly improved the content and presentation of this manuscript. FGM was supported in part by the National Science Foundation (CCF/CIF 1815971).

References

1. Abbe, E.: Community detection and stochastic block models: recent developments. J. Mach. Learn. Res. **18**(177), 1–86 (2018)
2. Airoldi, E.M., Costa, T.B., Chan, S.H.: Stochastic blockmodel approximation of a graphon: theory and consistent estimation. In: Advances in Neural Information Processing Systems, pp. 692–700 (2013)
3. Baldesi, L., Markopoulou, A., Buttsc, C.T.: Spectral graph forge: a framework for generating synthetic graphs with a target modularity. IEEE/ACM Trans. Netw. **27**(5), 2125–2136 (2019)
4. Chan, S., Airoldi, E.: A consistent histogram estimator for exchangeable graph models. In: International Conference on Machine Learning, pp. 208–216. PMLR (2014)
5. Franco Saldaña, D., Y, Yu., Feng, Y.: How many communities are there? J. Comput. Graph. Stat. **26**(1), 171–181 (2017)
6. Deng, S., Ling, S., Strohmer, T.: Strong consistency, graph Laplacians, and the stochastic block model. J. Mach. Learn. Res. **22**(117), 1–44 (2021)
7. Devriendt, K., Lambiotte, R., Van Mieghem, P.: Constructing Laplacian matrices with Soules vectors. arXiv preprint arXiv:1909.11282 (2019)
8. Elsner, L., Nabben, R., Neumann, M.: Orthogonal bases that lead to symmetric nonnegative matrices. Linear Algebra Appl. **271**(1–3), 323–343 (1998)
9. Eubanks, S., McDonald, J.J.: On a generalization of Soules bases. SIAM J. Matrix Anal. Appl. **31**(3), 1227–1234 (2010)
10. Ferguson, D., Meyer, F.G.: Theoretical analysis and computation of the sample Fréchet mean of sets of large graphs. Inf. Infer. **12**(3), 1347–1404 (2023)

11. Ferrer, M., Serratosa, F., Sanfeliu, A.: Synthesis of median spectral graph. In: Pattern Recognition and Image Analysis, pp. 139–146 (2005)
12. Löwe, M., Terveer, S.: Hitting times for random walks on the stochastic block model. arXiv preprint arXiv:2401.07896 pp. 1–26 (2024)
13. Soules, G.W.: Constructing symmetric nonnegative matrices. Linear Multilinear Algebra **13**(3), 241–251 (1983)
14. Meyer, F.G.: The spectral barycentre network (2025). https://github.com/francoismeyer/barycentre-network
15. Meyer, F.G., Shen, X.: Perturbation of the eigenvectors of the graph Laplacian: application to image denoising. Appl. Comput. Harmon. Anal. **36**(2), 326–334 (2014). https://doi.org/10.1016/j.acha.2013.06.004
16. Meyer, F.G.: When does the mean network capture the topology of a sample of networks? Front. Phys. **12**, 1–11 (2024). https://doi.org/10.3389/fphy.2024.1455988
17. Meyer, F.G.: The best soules basis for the estimation of a spectral barycentre network. arXiv preprint arXiv:2502.00038 (2025). https://arxiv.org/abs/2502.00038
18. Olhede, S.C., Wolfe, P.J.: Network histograms and universality of blockmodel approximation. PNAS **111**(41), 14722–14727 (2014)
19. Redko, I., Sebban, M., Habrard, A.: Non-negative matrix factorization meets time-inhomogeneous Markov chains. In: OPT2020 (2020)
20. Shen, X., Meyer, F.: Low-dimensional embedding of fMRI datasets. Neuroimage **41**(3), 886–902 (2008). https://doi.org/10.1016/j.neuroimage.2008.02.051
21. Shine, A., Kempe, D.: Generative graph models based on Laplacian spectra? In: The World Wide Web Conference, pp. 1691–1701. ACM (2019)
22. Thiele, C.M., Villemoes, L.F.: A fast algorithm for adapted time-frequency tilings. Appl. Comput. Harmon. Anal. **3**(2), 91–99 (1996)
23. Van Dam, E.R., Haemers, W.H.: Developments on spectral characterizations of graphs. Discret. Math. **309**(3), 576–586 (2009)
24. White, D., Wilson, R.C.: Spectral generative models for graphs. In: ICIAP 2007, pp. 35–42 (2007)
25. Wills, P., Meyer, F.G.: Metrics for graph comparison: a practitioner's guide. PLoS ONE **15**(2), 1–54 (2020). https://doi.org/10.1371/journal.pone.0228728
26. Yan, B., Sarkar, P., Cheng, X.: Provable estimation of the number of blocks in block models. In: ICAIS, pp. 1185–1194 (2018)

The Multilayer Artificial Benchmark for Community Detection (mABCD)

Piotr Bródka[1(✉)], Michał Czuba[1], Bogumił Kamiński[2], Łukasz Kraiński[2], Paweł Prałat[3], and François Théberge[4]

[1] Department of Artificial Intelligence, Wrocław University of Science and Technology, Wrocław, Poland
`{piotr.brodka,michal.czuba}@pwr.edu.pl`

[2] Decision Analysis and Support Unit, SGH Warsaw School of Economics, Warsaw, Poland
`{bkamins,lkrain}@sgh.waw.pl`

[3] Department of Mathematics, Toronto Metropolitan University, Toronto, Canada
`pralat@torontomu.ca`

[4] Tutte Institute for Mathematics and Computing, Ottawa, Canada
`theberge@ieee.org`

Abstract. The Artificial Benchmark for Community Detection (**ABCD**) is a random graph model that incorporates community structure and follows a power-law distribution for both node degrees and community sizes. It produces graphs similar to the well-known LFR model but is faster, more interpretable, and analytically tractable. In this paper, we build on the core principles of **ABCD** to introduce a **mABCD**, a new variant designed for multilayer networks.

Keywords: ABCD · community detection · multilayer networks · benchmark models

1 Introduction

Community structure is a key feature of real-world networks, revealing their internal organization [15,20]. In social networks, it reflects shared interests; in citation networks, it groups related papers; and in the Web graph, it connects pages on similar topics. Identifying communities helps in understanding network structure.

However, datasets with ground-truth communities are scarce, necessitating synthetic graph models for benchmarking clustering algorithms. The **LFR** model [26,27] is widely used to generate networks with community structures while allowing heterogeneity in node degrees and community sizes.

A more recent alternative, the **ABCD** (**A**rtificial **B**enchmark for **C**ommunity **D**etection) model [19], along with its fast multi-threaded version **ABCDe** [23], provides comparable properties to **LFR** but is faster and more flexible. It enables smooth transitions between disjoint communities and random graphs and is theoretically easier to analyze [18]. Notably, it exhibits self-similar degree distributions [3] and has been extended to handle outliers (**ABCD+o**) [21] and hypergraphs (**h–ABCD**) [22].

The **ABCD** model is gaining traction among both practitioners and researchers. For instance, [1] employs **Adjusted Mutual Information (AMI)** to compare 30 community detection algorithms using **LFR** and **ABCD**, highlighting **ABCD**'s scalability and improved parameter control. Given its flexibility, this paper extends **ABCD** to generate multilayer networks.

Multilayer network [14,25] is a type of complex network that has gathered a lot of attention over the last decade. Such networks allow us to capture multiple different relations between nodes, thus better reflecting the complexity of real-world interactions. Shortly after the emergence of the multilayer network concept, the research on community detection in those networks begun [6,10,35] (a detailed description can be found in the survey paper [28]). This, and the fact that publicly available real-world datasets with known community structures for multilayer networks are even less common, generated the need for synthetic multilayer networks with known community structures.

The first such models were extensions of the existing models for simple single-layer networks such as **mLFR** [8], which is an extension of the **LFR** model mentioned above. Another model was proposed in [4]. It is designed to generate networks with various kinds of meso-structures, with the community being an example of such a structure. Finally, the authors of the survey paper on community detection in multilayer networks [28] proposed yet another model that allows the generation of a simple community structure and a network with edges based on that structure. This model allowed them to compare several existing community detection algorithms and was partially included in the *multinet* library [29] for analyzing and mining multilayer networks.

While these models provide valuable tools for benchmarking community detection algorithms, each has its trade-offs. Magnani et al.'s [28] approach prioritizes simplicity and algorithm comparison but sacrifices scalability and complexity. Bazzi et al. [4] offer flexibility and mesoscale modelling but face challenges with parameter complexity and computational efficiency. Finally, the **mLFR** model [8] focuses on structural realism but lacks support for heterogeneous and strongly correlated inter-layer properties. These limitations underscore the ongoing need for more advanced models to replicate the multifaceted nature of real-world multilayer networks.

In this paper, we use the underlying ingredients of the **ABCD** model and introduce its variant for multilayer networks, **mABCD**. The paper is structured as follows. First, we introduce the notation and terminology for multilayer networks—see Sect. 2. In particular, we introduce various correlation measures between the layers. In Sect. 3, we formally define the **mABCD** model. Conclusions and future directions are presented in Sect. 4.

2 Multilayer Networks

In this section, we introduce the standard notation used and properties of interest for multilayer networks [14,25]. For a given $n \in \mathbb{N} := \{1, 2, \ldots\}$, we use $[n]$ to denote the set consisting of the first n natural numbers, that is, $[n] := \{1, 2, \ldots, n\}$. We define a multilayer network as a quadruple $M = ([n], [\ell], V = [n] \times [\ell], E)$, where

- $[n]$ is a set of n actors (for example, users of various social networking sites),
- $[\ell]$ is a set of layers (for example, different social networking platforms, such as LinkedIn, Facebook and Instagram, on which actors interact with each other),
- $V \subseteq [n] \times [\ell]$ a set of nodes (vertices); node $v = (a, \ell_i) \in V$ represents an actor a in layer ℓ_i,
- E is a set of (undirected) edges between nodes; if $e = v_1 v_2 \in E$ with $v_1 = (a_1, \ell_1) \in V$ and $v_2 = (a_2, \ell_2) \in V$, then $\ell_1 = \ell_2$, that is, edges occur only within layers.

Note that not every actor needs to be present on all layers. For simplicity, in our model, we assume that each layer has exactly n nodes associated with all actors. Actors not engaging with a given layer (we will call them inactive) will be associated with isolated nodes (nodes of degree zero).

2.1 Correlations Between Layers

There are no edges between nodes in different layers, but in most real-world multilayered networks, layers are clearly *not* independently generated. Each actor is associated with ℓ nodes, one in each layer, and there are some highly non-trivial correlations between edges across layers. For example, active users on one social media platform are often also active on another one [16]. This creates correlations between degree distributions across layers. Communities that are naturally formed in various layers often depend on the properties of the associated actors. For example, users interested in soccer might group together on Instagram and on Facebook. As a result, partitions of nodes into communities (associated with different layers) are often correlated. Finally, interactions between actors in one layer might increase their chances of interacting in another layer, yielding correlations at the level of edges.

Below, we briefly summarize how we measure these three types of correlations mentioned above. The first two measures are standard, and their detailed description can be found, for example, in [9,20].

Correlations Between Nodes Degrees in Various Layers. We will use **Kendall rank correlation coefficient** τ [24] to measure correlations between sequences of node degrees in two different layers. It is a nonparametric measure of the ordinal association between two measured quantities: the similarity of the orderings of the data when ranked by each of the quantities (in our application, the degree sequences). The Kendall correlation between two variables ranges from -1 to 1. It is large when observations have a similar rank between the two variables and is small when observations have a dissimilar rank between the two variables.

Specifically, we will use the "tau-b" statistic, which is adjusted to handle ties. If an actor is inactive in one of the two layers we compare against each other, then we simply ignore the two nodes corresponding to this actor. As a result, the degree sequences are always of the same length.

Correlations Between Partitions in Various Layers. The **adjusted mutual information (AMI)**, a variation of **mutual information (MI)**, is a common way to compare partitions of the same set [20, 36]. Usually, one may want to compare the partitions returned by some clustering algorithms. In our present context, we may want to compare partitions into ground-truth communities from two different layers. The **AMI** takes a value of 1 when the two partitions are identical and 0 when the **MI** between two partitions equals the value expected due to chance alone. Actors, that are inactive in at least one of the two layers we compare against each other, are ignored so that a comparison of partitions is made on the same set of actors.

Correlations Between Edges in Various Layers. To measure correlations between edges in different layers, we define $\mathbf{R}$, a $\ell \times \ell$ matrix in which elements $r_{i,j} \in [0, 1]$ ($i, j \in [\ell]$) capture correlation between edges present in layers i and j. For any $i, j \in [\ell]$ with $i < j$, let

$$E_i^j = \{a_1 a_2 : (a_1, i)(a_2, i) \in E \wedge a_1, a_2 \in [n] \text{ are active in layers } i \text{ and } j\}, \quad (1)$$

be the set of edges that are present in layer i, involving actors that are also active in layer j. Note that in the definition of E_i^j, edges are defined over actors that are active in both layers, so that we can perform set operations on edges between layers. Entries $r_{i,j}$ in $\mathbf{R}$ are computed using the following formula:

$$r_{i,j} = \frac{|E_i^j \cap E_j^i|}{\min\{|E_i^j|, |E_j^i|\}}.$$

If $\min\{|E_i^j|, |E_j^i|\} = 0$, then we leave $r_{i,j}$ undefined; in the implementation, NaN value is produced.

Note that the definition of $\mathbf{R}$ implies that $r_{i,i} = 1$ for any $i \in [\ell]$ and $r_{i,j} = r_{j,i}$ for $1 \le i < j \le \ell$. The maximum value of 1 is attained when edges in one of the layers form a subset of edges in the other layer. The minimum value of 0 is attained when the two sets of edges in the corresponding layers are completely disjoint. As a result, r_{ij} aims to capture correlations between individual edges, but it is not normalized as, for example, the Kendall rank correlation coefficient τ. The coefficient τ ranges from -1 to 1, corresponding to the two extremes, and 0 corresponds to a neutral case. Graphs associated with the layers are sparse, but one layer might have substantially more edges than the other. Hence, r_{ij} is convenient, but it does not have a natural interpretation as τ. Finally, let us mention that one can easily update the value of r_{ij} when some small operations are applied to either layer i or j. It will become handy when such operations must be performed on our synthetic model to converge to the desired correlation matrix **R**.

2.2 Examples of Multilayer Networks

Before constructing the model, we first examined whether and how degrees, edges, and partitions correlate across layers in real-world networks. Existing research [9,28] has yet to provide a definitive answer. To fill this gap, we analyzed eight real-world networks from diverse domains, each differing in actors, nodes, edges, and layers. Table 1 summarizes their structural characteristics, while Fig. 1 presents correlation matrices for node degrees, partitions, and edges (as defined in Sect. 2.1) for one of the analyzed networks.

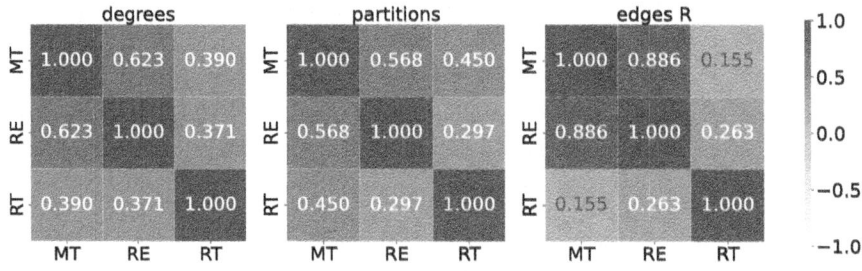

Fig. 1. Node degrees, partitions, and edge correlations between layers of cannes network (MT- mentions, RE - replies, and RT - retweets).

Our analysis revealed no consistent patterns across networks-degrees, edges, and partitions were sometimes correlated and sometimes independent. In one network, two layers might be correlated, while another layer remained unaligned. To accommodate this variability, we designed **mABCD** for flexibility: advanced users can customize distributions and correlations between layers, while default parameters, such as a power-law degree distribution, provide guidance for less experienced users.

Table 1. Networks used in experiments with their basic parameters shortlisted.

Name	Layers	Actors	Nodes	Edges	Note
arxiv	13	14,065	26,796	59,026	Coauthorship network from articles published on the "arXiv" [13]
aucs	5	61	224	620	A graph of interactions between employees of Aarhus University, Department of Computer Science [33]
cannes	3	438,537	659,951	974,743	A network of interactions between Twitter users [30]
ckmp	3	241	674	1,370	A network depicting diffusion of innovations among physicians [12]
eutr-A	37	417	2,034	3,588	The European air transportation network [11]
l2-course	2	41	82	297	A network of interactions between U.S. students learning Arabic [32]
lazega	3	71	212	1,659	A network of interactions between staff of a law corporation [34]
timik	3	61,702	102,247	881,676	A graph of interactions between users of the virtual world platform [17]

3 The mABCD Model

In this section, we introduce a variant of the **ABCD** model that produces a synthetic collection of graphs that form a multilayer structure, **mABCD**.

3.1 Power-Law Distribution

Power-law distributions will be used to generate both the degree sequence and community sizes so let us formally define it. For given parameters $\gamma \in (0, \infty)$, $\delta, \Delta \in \mathbb{N}$ with $\delta \leq \Delta$, we define a truncated power-law distribution $\mathcal{P}(\gamma, \delta, \Delta)$ as follows. For $X \sim \mathcal{P}(\gamma, \delta, \Delta)$ and for $k \in \mathbb{N}$ with $\delta \leq k \leq \Delta$,

$$\mathbb{P}(X = k) = \frac{\int_k^{k+1} x^{-\gamma}\, dx}{\int_\delta^{\Delta+1} x^{-\gamma}\, dx}.$$

3.2 The Configuration Model

The well-known configuration model is an important ingredient of the generation process, so let us formally define it here. Suppose then that our goal is to create a graph on n nodes with a given degree sequence $\mathbf{d} := (d_i, i \in [n])$, where $\mathbf{d}$ is a sequence of non-negative integers such that $m := \sum_{i \in [n]} d_i$ is even. We define a random multi-graph CM($\mathbf{d}$) with a given degree sequence known as the **configuration model** (sometimes called the **pairing model**), which was first introduced by Bollobás [7]. (See [5, 37, 38] for related models and results.)

We start by labelling nodes as $[n]$ and, for each $i \in [n]$, endowing node i with d_i half-edges. We then iteratively choose two unpaired half-edges uniformly at random (from the set of pairs of remaining half-edges) and pair them together to form an edge. We iterate until all half-edges have been paired. This process yields $G_n \sim \text{CM}(\mathbf{d})$, where G_n is allowed self-loops and multi-edges and thus G_n is a multi-graph.

3.3 Parameters of the mABCD Model

The **mABCD** model is governed by the following parameters. The first family of parameters is responsible for a few global properties of the model.

Parameter	Range	Description
n	$\mathbb{N}$	Number of actors
ℓ	$\mathbb{N}$	Number of layers
$\mathbf{R}$	$[0,1]^{\ell \times \ell}$	Correlation between edges
d	$\mathbb{N}$	Dimensionality of reference layer

Actors will be associated with *labels* from the set $[n]$. These labels will affect the degrees of actors. Each actor $a \in [n]$ will be associated with ℓ nodes, $v_i = (a, i)$ with $i \in [\ell]$, one for each of the ℓ layers. Moreover, each actor will be associated with a vector in $\mathbb{R}^d$ representing their features. We will refer to these vectors as vectors in the *reference layer*. This reference layer will affect the process of generating partitions into communities in various layers.

The second family of parameters is responsible for various properties that are specific for each of the ℓ layers; subscripts $i \in [\ell]$ indicate that the corresponding parameters shape the ith layer. In particular, the set of parameters ξ_i, $i \in [\ell]$, will control the level of noise, that is, the fraction of edges in layer i that are between nodes from two different communities.

Parameter	Range	Description
q_i	$(0,1]$	Fraction of active actors
τ_i	$[-1,1]$	Correlation coefficient between degrees and labels
r_i	$[0,1]$	Correlation between communities and reference layer
γ_i	$(2,3)$	Exponent of power-law degree distribution
δ_i	$\mathbb{N}$	Min degree at least δ_i
Δ_i	$\mathbb{N}\ (1 \le \delta_i \le \Delta_i < n)$	Max degree at most Δ_i
β_i	$(1,2)$	Exponent of power-law community size distribution
s_i	$\mathbb{N}$	Min community size at least s_i
S_i	$\mathbb{N}\ (\delta < s_i \le S_i \le n)$	Max community size at most S_i
ξ_i	$(0,1)$	Level of noise

The suggested range of values for parameters γ and β are chosen according to experimental values commonly observed in complex networks [2,31].

3.4 The mABCD Construction

We will use $\mathcal{A}$ for the distribution of graphs (layers) generated by the following 6-phase construction process. The model generates ℓ graphs; graph $G_n^i = ([n] \times \{i\}, E^i)$, $i \in [\ell]$, is the graph representing the ith layer. Once they are generated, we simply take $V = \bigcup_{i \in [\ell]}([n] \times \{i\})$ and $E = \bigcup_{i \in [\ell]} E^i$.

Phase 1: Selecting Active Nodes. As mentioned above, in a multilayer network, not all of the actors are active in all the layers. Actor $a \in [n]$ is *active* in layer i with probability q_i, independently for each $i \in [\ell]$ and all other actors. If an actor is not active in a given layer, then it will be represented by an "artificial" *inactive* node. We use N^i to denote the number of active nodes (and actors) in layer i. (Clearly, N^i is a random variable with expectation $q_i n$.) For convenience, we will keep inactive nodes as part of the corresponding graphs, but one may think of them as being removed from a given layer.

Phase 2: Creating Degree Sequences. The degree sequences for all of the ℓ layers are generated independently so we may concentrate on a given layer $i \in [\ell]$. We ensure that the degree sequence satisfies (a) a power-law with parameter γ_i, (b) a minimum value of at least δ_i, and (c) a maximum value of at most Δ_i.

Inactive nodes (representing actors not present in particular layer) are easy to deal with, they simply have degree zero. The remaining N^i degrees are i.i.d. samples from the distribution $\mathcal{P}(\gamma_i, \delta_i, \Delta_i)$. We use $\mathbf{d}_n^i = (d_v^i, v \in [n])$ for the generated degree sequence of G_n^i with $d_1^i \geq d_2^i \geq \cdots \geq d_n^i$; $\mathbf{d}_{N^i}^i$ is a degree subsequence of active nodes. Finally, to ensure that $\sum_{v \in [n]} d_v^i$ is even, we decrease d_1^i by 1 if necessary; we relabel as needed to ensure that $d_1^i \geq d_2^i \geq \cdots \geq d_n^i$.

Parameter $\tau_i \in [-1, 1]$ controls how degrees of the nodes are correlated with labels of the associated actors (recall that node (a, i) in layer i is associated with an actor with label a). In one important case, namely, when $\tau_i = 0$, there is no correlation at all and the degree sequence $\mathbf{d}_{N^i}^i$ is assigned randomly to the N_i active nodes. When $\tau_i = 1$, the order of active nodes with respect to their labels is the same as the order with respect to their degrees; the largest degree node is first. In other words, if nodes $(a_1, i), (a_2, i)$ with $1 \leq a_1 < a_2 \leq n$ are active, then $\deg^i(a_1) \geq \deg^i(a_2)$, where $\deg^i(a_1)$ is the degree of node (a_1, i). Similarly, if $\tau_i = -1$, then the order of active nodes with respect to their labels is also consistent with the order with respect to their degrees but this time the last node is of the largest degree. Since τ_i's could be different for different layers, one node could have large degrees in some layers but small ones in some other ones.

To achieve the desired property, each active node (a, i) independently generates a normally distributed random variable $X_a = N(a/n, \sigma_i)$, where the variance σ_i is a specific function of τ_i. (Recall that we concentrate on a given layer $i \in [\ell]$. For convenience, we simplify the notation and stop referencing to layer i

in notation such as X_a. Still, there are many independent random variables for each active node (a, i) associated with actor a.) We sort active nodes in increasing order of their values of X_a and assign the degree sequence accordingly; that is, node (a, i) gets degree d_r^i, where $r \in [N_i]$ is the rank of X_a. In particular, the node with the smallest value of X_a gets assigned the largest degree, namely, d_1^i. Note that if $\sigma_i = 0$, then $X_a = a/n$ (deterministically), and so we recover the perfect correlation between the degrees and the labels ($\tau_i = 1$). On the other hand, if $\sigma_i \to \infty$, then the order of nodes is perfectly random (with uniform distribution), so we recover the other desired extreme ($\tau_i = 0$).

Function $\sigma_i : [0, 1] \to [0, \infty)$ is empirically approximated so that the variance $\sigma_i = \sigma_i(\rho_i)$ yields the Kendall rank correlation close to $\tau_i \in [0, 1]$ between the ordering generated by the ranks of X_a and the labels a associated with corresponding actors that are active in layer i. Twenty degree distributions are independently generated (with the same σ and different random seeds), and the one with the correlation coefficient that is the closest to the desired value of τ_i is kept. To deal with negative correlations $\tau_i \in [-1, 0)$, we simply "flip" the order generated for $|\tau_i|$.

Phase 3: Creating Communities. Our next goal is to create community structure in each layer of the **mABCD** model. When we construct a community, we assign a number of nodes to said community equal to its size. Initially, the communities form empty graphs. Then, in later phases we handle the construction of edges using the degree sequence established in Phase 2.

Similarly to the process of generating the degree sequences, the sequence of community sizes are generated independently, ensuring that the distribution for a given layer $i \in [\ell]$, satisfy (a) a power-law with parameter β_i, (b) a minimum value of s_i, and (c) a maximum value of S_i. In addition, we also require that the sum of community sizes is exactly n. Specifically, inactive nodes (if there are any) form their own community, namely, C_0^i. Other communities are generated with sizes determined independently by the distribution $\mathcal{P}(\beta_i, s_i, S_i)$. We generate communities until their collective size is at least n. If the sum of community sizes at this moment is $n + x$ with $x > 0$, then we perform one of two actions: if the last added community has a size at least $x + s_i$, then we reduce its size by x. Otherwise (that is, if its size is $c < x + s_i$), then we delete this community, select $c - x$ old communities and increase their sizes by 1.

Now, given that the sequences of community sizes are already determined (for all layers), it is time to assign nodes to communities. To allow communities to be correlated with each other, we first create a latent *reference* layer that will guide the process of assigning nodes to specific communities across all layers. One may think of this auxiliary layer as properties of actors (such as people's age, education, geographic location, beliefs, etc.) shaping different layers (for example, various social media platforms). This single reference layer will be used for all ℓ layers. In this reference layer, each actor $a \in [n]$ gets assigned a random vector in $\mathbb{R}^d$ (by default, $d = 2$) that is taken independently and uniformly at random from the ball of radius one centred at $\mathbf{0} = (0, 0, \ldots, 0)$.

Let us now concentrate on a given layer $i \in [\ell]$. Recall that the community sizes have already been generated. We write L^i for the (random) number of regular communities in layer i partitioning the set of active nodes in this layer and use $\mathbf{c}^i = (c^i_j, j \in \{0\} \cup [L^i])$ for the corresponding sequence of community sizes. Recall that inactive nodes (representing actors not present in a particular layer) form their own community (namely, C^i_0) so $c^i_0 = n - N^i$ is the number of inactive nodes in layer i. Let R be the set of active nodes. We assign nodes to communities, dealing with one community at a time in a random order. When community C^i_j is formed (for some $j \in [L^i]$), we first select a node from R that is at the largest distance from the center $\mathbf{0}$ (in the reference layer). This node, together with its $c^i_j - 1$ nearest neighbours in R, are put to C^i_j. We remove C^i_j from R and move on to the next community.

The above strategy creates a partition of nodes that is highly correlated with the geometric locations of nodes in the reference layer; nodes that are close to each other in the reference layer are often in the same community—see Fig. 2 for an example. To reduce the correlation strength (modelled by the parameter $r_i \in [0,1]$), we perform the following procedure. Each active node independently leaves its own community with probability $1 - r_i$, freeing a spot in this community. All the nodes that left are then put back randomly to any available spot (which typically is in a community that this node was *not* originally in). Note that in the extreme case, when $r_i = 0$, the resulting partition does not depend on the reference layer at all, so there is no correlation. We write $\mathbf{C}^i_n = (C^i_j, j \in \{0\} \cup [L^i])$ for the generated collection of communities in G^i_n. Again, let us stress the fact that $\mathbf{C}^i_n$ is a random partition of $[n]$ of random size $L^i + 1$.

Finally, note that in the above process of assigning nodes to communities, as opposed to the original **ABCD** model, we ignore the degree of nodes. Indeed, the original *ABCD* model tries to make sure that large degree nodes are not assigned to small communities. In **mABCD**, there are many layers and non-trivial correlations between partitions into communities and degree sequences between layers. In a hypothetical extreme situation, it might happen that each node belongs to some small community in some layer. Hence, the **mABCD** model does not try to prevent such unavoidable situations and will resolve potential issues later (see Phase 5).

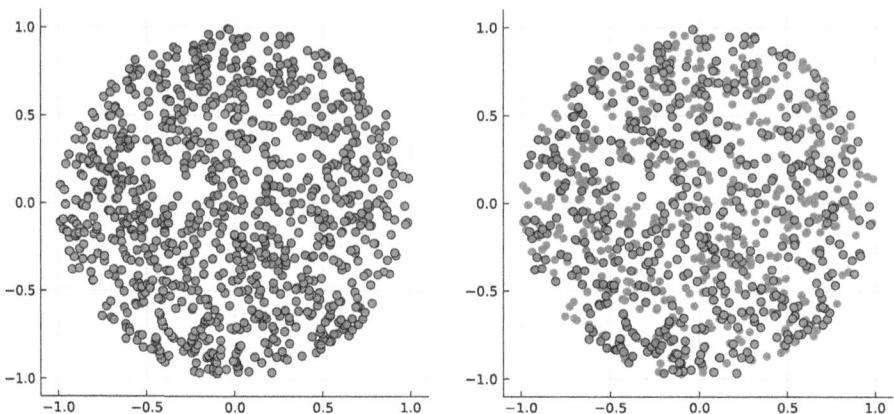

Fig. 2. Two partitions generated based on the same reference layer with $n = 1{,}000$ nodes: (left) $q_1 = 1$ (all nodes active), $S_1 = 32$, $s_1 = 16$, $\beta_1 = 1.5$, (right): $q_2 = 0.5$ (50% nodes active), $S_2 = 50$, $s_2 = 25$, $\beta_2 = 1.5$.

Phase 4: Creating Edges. Now, it is time to form edges in **mABCD**. It will be done in the next three phases, Phases 4–6. Phases 4 and 5 will independently generate ℓ graphs G_n^i, $i \in [\ell]$, for each of the ℓ layers whereas Phase 6 will make sure that edges across various layers are correlated, if needed. We may then concentrate on a given layer $i \in [\ell]$.

At this point G_n^i contains n nodes labelled as (a, i), $a \in [n]$, partitioned by the communities $\mathbf{C}_n^i$, with node (a, i) containing $\deg^i(a)$ unpaired half-edges. Firstly, for each $a \in [n]$, we split the $\deg^i(a)$ half-edges of (a, i) into two distinct groups, which we call *community* half-edges and *background* half-edges. For $a \in \mathbb{Z}$ and $b \in [0, 1)$ define the random variable $\lfloor a + b \rceil$ as

$$\lfloor a + b \rceil = \begin{cases} a & \text{with probability } 1 - b, \text{ and} \\ a + 1 & \text{with probability } b. \end{cases}$$

(Note that $\mathbb{E}\left[\lfloor a + b \rceil\right] = a(1 - b) + (a + 1)b = a + b$.) Now define $Y_a := \lfloor (1 - \xi) \deg^i(a) \rceil$ and $Z_a := \deg^i(a) - Y_a$ (note that Y_a and Z_a are random variables with $\mathbb{E}[Y_a] = (1 - \xi) \deg^i(a)$ and $\mathbb{E}[Z_a] = \xi \deg^i(a)$ and since we generate each layer separately they are different for each layer) and, for all $a \in [n]$, split the $\deg^i(a)$ half-edges of (a, i) into Y_a community half-edges and Z_a background half-edges. Next, for all $j \in [L^i]$, construct the *community graph* $G_{n,j}^i$ as per the configuration model on node set C_j^i and degree sequence $(Y_a, a \in C_j^i)$. Note that C_0 consists of inactive nodes which, by design, have degree zero. Hence, there is no need to do anything with them. Finally, construct the *background graph* $G_{n,0}^i$ as per the configuration model on node set $[n]$ and degree sequence $(Z_a, a \in [n])$. In the event that the sum of degrees in a community is odd, we pick a maximum degree node $(a, i))$ in said community and replace Y_a with $Y_a + 1$ and Z_a with $Z_a - 1$. Note that $G_{n,j}^i$ is a graph, and C_j^i is the set of nodes in this graph;

we refer to C^i_j as a *community* and $G^i_{n,j}$ as a *community graph*. Note also that $G^i_n = \bigcup_{0 \le j \le L^i} G^i_{n,j}$.

Phase 5: Rewiring Self-loops and Multi-edges. We continue concentrating on a given layer $i \in [\ell]$. Note that, although we are calling $G^i_{n,j}$ ($j \in \{0\} \cup [L^i]$) *graphs*, they are in fact *multi-graphs* at the end of Phase 4. To ensure that G^i_n is simple, we perform a series of *rewirings* in G^i_n. A rewiring takes two edges as input, splits them into four half-edges, and creates two new edges distinct from the input. We first rewire each community graph $G^i_{n,j}$, $j \in [L^i]$, independently as follows.

1. For each edge $e \in E(G^i_{n,j})$ that is either a loop or contributes to a multi-edge, we add e to a *recycle* list that is assigned to $G^i_{n,j}$.
2. We shuffle the *recycle* list and, for each edge e in the list, we choose another edge e' uniformly from $E(G^i_{n,j}) \setminus \{e\}$ (not necessarily in the *recycle* list) and attempt to rewire these two edges. We save the result only if the rewiring does not lead to any further self-loops or multi-edges, otherwise we give up. In either case, we then move to the next edge in the *recycle* list.
3. After we attempt to rewire every edge in the *recycle* list, we check to see if the new *recycle* list is smaller. If yes, we repeat step 2 with the new list. If no, we give up and move all of the "bad" edges from the community graph to the background graph.

We then rewire the background graph $G^i_{n,0}$ in the same way as the community graphs, with the slight variation that we also add edge e to *recycle* if e forms a multi-edge with an edge in a community graph or, as mentioned previously, if e was moved to the background graph as a result of giving up during the rewiring phase of its community graph. At the end of Phase 5, we have a simple graph G^i_n representing the i-th layer of a multilayer network.

Phase 6: Correlations Between Edges in Various Layers. During this last phase, we continue performing a series of rewiring (in batches) with the goal of creating a multilayer network with the correlations between edges in various layers (as defined in Subsect. 2.1) to be as close to the desired matrix $\mathbf{R}$ (provided as one of the parameters of the model) as possible. It is important to highlight the fact that during this phase, not only do the degrees of the involved nodes not change, but the community degrees stay the same (as well as the background ones). Hence, in particular, the level of noise stays the same.

We run t independent *batches* of operations (by default, $t = 100$). Before every batch, we re-compute the (empirical) correlation matrix $\hat{\mathbf{R}}$ for the current multilayer network $(G^i_n : i \in [\ell])$ and compare it with the desired matrix $\mathbf{R}$. We select an entry ij at random with the probability proportional to the discrepancy between the empirical and the desired values. In other words, we select a pair (i,j) ($1 \le i < j \le \ell$) with probability

$$p_{ij} = \frac{|\hat{r}_{ij} - r_{ij}|}{\sum_{1 \le r < s \le \ell} |\hat{r}_{rs} - r_{rs}|}.$$

We attempt to rewire $\lceil \epsilon \min\{|E_i^j|, |E_j^i|\} \rceil$ of edges in each batch with the goal to bring $\hat{r}_{ij}$ closer to r_{ij} (by default, $\epsilon = 0.05$). Recall that E_i^j can be viewed as the set of edges in layer i that are between actors that are active in layer j (and, trivially, also active in layer i since inactive actors form isolated nodes), see (1).

Suppose first that $\hat{r}_{ij} < r_{ij}$, that is, the correlation between layer i and layer j is smaller than what we wished for. Each of the attempts does the following. Randomly select one of the two graphs, G_n^i or G_n^j, and call it *primary*. Then, pick a random edge uv from the primary graph between actors that are active in both layers. Our goal is to try to introduce edge uv in the other graph (call it *secondary*) unless it is already there, in which case we simply finish this attempt prematurely. If u and v belong to one of the communities in the secondary graph (say to the community C), then we take u' to be a random neighbour of u in C (if there are any), take v' to be a random neighbour of v in C (again, if there are any). If u, v, u', v' are four distinct nodes and there is no edge $u'v'$ in the secondary graph, then we remove the two edges uu' and vv' and introduce two new edges uv and $u'v'$. If anything goes wrong, then we simply finish prematurely and move on to another attempt. If u and v are from two different communities in the secondary graph, then the procedure is exactly the same, but this time, our goal is to select four nodes, each from a different community. Specifically, we try to pick a random neighbour u' of u outside of the communities u or v belong to. Then, we try to pick a random neighbour v' of v outside of the communities u, v, or u' belong to. If the four selected nodes are different and there is no edge $u'v'$ in the secondary graph, we do the rewiring.

Suppose now that $\hat{r}_{ij} > r_{ij}$, that is, the correlation between layer i and layer j is larger than what we wished for. As before, during each attempt, we randomly make one of the two graphs, G_n^i, G_n^j, to be *primary* and the second one to be *secondary*. Then, pick a random edge uv from the intersection of the two graphs. Our goal is to try to remove edge uv from the secondary graph. If u and v belong to one of the communities in the secondary graph (say to the community C), then we take a random edge $u'v'$ from C. If u, v, u', v' are four distinct nodes and there are no edges uu' nor vv' in the secondary graph, then we remove the two edges uv and $u'v'$ and introduce two new edges uu' and vv'. As before, if anything goes wrong, then we simply finish prematurely and move on to another attempt. If u and v are from two different communities in the secondary graph, then we pick a random edge $u'v'$ from the secondary graph with the property that all four nodes belong to different communities. We try to rewire the two edges, making sure that no multi-edges get created.

The goal of the sequence of t batches is to bring the (empirical) correlation matrix $\hat{\mathbf{R}}$ closer to the desired matrix $\mathbf{R}$. Unfortunately, fixing one entry of $\mathbf{R}$ may affect the other entries. Hence, it is not guaranteed that the best solution is found after exactly t batches. To take this into account, we track the quality of the multilayer networks $(G_n^i : i \in [\ell])$ at the beginning of each bath (via L_2 norm between $\hat{\mathbf{R}}$ and $\mathbf{R}$) and the final network is one of the t networks that performed best.

4 Conclusions and Future Directions

In this paper, we introduced **mABCD**, an extension of the **ABCD** model designed for multilayer networks. This new benchmark model provides a flexible and scalable framework for generating synthetic multilayer networks with ground-truth communities while maintaining desirable properties such as power-law degree and community size distributions.

A key contribution of **mABCD** is its ability to control correlations between layers, including node degrees, community structures, and edge connections. Unlike some of the existing multilayer benchmarks, which often impose rigid structural assumptions, **mABCD** allows for a smooth transition between independent layers and highly correlated structures. This makes it particularly suitable for evaluating community detection algorithms across different network configurations.

Another advantage of **mABCD** is its computational efficiency. By leveraging the foundational principles of **ABCD**, it remains faster and more interpretable than other state-of-the-art models like multilayerGM [4] while enabling users to fine-tune network properties to better reflect real-world multilayer interactions.

Our empirical analysis of real-world multilayer networks highlighted the high variability in inter-layer correlations, underscoring the need for a flexible synthetic model. **mABCD** directly addresses this need by allowing users to either fully customize correlation structures or rely on default settings that mimic observed patterns.

With its scalability, adaptability, and theoretical tractability, **mABCD** is a powerful tool for researchers and practitioners working on community detection in multilayer networks. We believe that this benchmark will facilitate more robust evaluations and drive further advancements in the field.

The future work which we aim to present in the follow-up journal paper will include a deeper theoretical analysis of its properties. Initial experiments confirm that **mABCD** effectively captures the desired correlations between degrees, community structures, and edges across layers. However, a more rigorous study of these relationships, including their behaviour and sensitivity to parameter variations, would enhance our understanding of the model's capabilities and limitations. Another area that needs deeper evaluation is correlation control between layers. Our current approach provides flexibility in tuning inter-layer dependencies, but we would like to empirically test if correlations between degrees, partitions, and edges always match the desired values and, if not, what are the tradeoffs between exact matching and model efficiency (generation speed). The computational efficiency of **mABCD** is the next property we would like to evaluate to see if the model could be optimized. For example, optimizing the correlation adjustment phase (Phase 6) could significantly reduce execution time for large networks. Finally, we would like to explore **mABCD** in other applications than community detection. For example, using **mABCD** to generate networks with different topologies (e.g., in terms of correlations between layers or weak/strong community structure) to evaluate how network topology can affect various dynamic processes on those networks. For example, assessing the impact of community structure on information diffusion, epidemic spreading, or opinion dynamics could provide valuable insights into real-world multilayer networks.

Acknowledgments. This research was partially supported by the National Science Centre, Poland, under Grant no. 2022/45/B/ST6/04145 (https://multispread.pwr.edu.pl/), the Polish Ministry of Education and Science within the programme "International Projects Co-Funded", and the EU under the Horizon Europe, grant no. 101086321 OMINO. Views and opinions expressed are, however, those of the authors only and do not necessarily reflect those of the National Science Centre, Polish Ministry of Education and Science, EU or the European Research Executive Agency.

Disclosure of Interests. The authors have no competing interests to declare that are relevant to the content of this article.

Code and Data. The algorithm is implemented in Julia programming language. Source code and installation instructions are available on mABCD GitHub repository (https://github.com/KrainskiL/MLNABCDGraphGenerator.jl).

References

1. Aref, S., Chheda, H., Mostajabdaveh, M.: The Bayan algorithm: detecting communities in networks through exact and approximate optimization of modularity. arXiv preprint arXiv:2209.04562 (2022)
2. Barabási, A.-L.: Network Science. Cambridge University Press (2016). http://barabasi.com/networksciencebook/
3. Barrett, J., Kamiński, B., Prałat, P., Théberge, F.: Self-similarity of communities of the ABCD model. Theor. Comput. Sci. **1026**, 115012 (2025)
4. Bazzi, M., Jeub, L.G.S., Arenas, A., Howison, S.D., Porter, M.A.: A framework for the construction of generative models for mesoscale structure in multilayer networks. Phys. Rev. Res. **2**(2), 023100 (2020)
5. Bender, E.A., Rodney Canfield, E.: The asymptotic number of labeled graphs with given degree sequences. J. Comb. Theory Ser. A **24**(3), 296–307 (1978)
6. Berlingerio, M., Coscia, M., Giannotti, F.: Finding and characterizing communities in multidimensional networks. In: 2011 International Conference on Advances in Social Networks Analysis and Mining, pp. 490–494. IEEE (2011)
7. Bollobás, B.: A probabilistic proof of an asymptotic formula for the number of labelled regular graphs. Eur. J. Comb. **1**(4), 311–316 (1980)
8. Bródka, P.: A method for group extraction and analysis in multilayer social networks. arXiv preprint arXiv:1612.02377 (2016)
9. Bródka, P., Chmiel, A., Magnani, M., Ragozini, G.: Quantifying layer similarity in multiplex networks: a systematic study. R. Soc. Open Sci. **5**(8), 171747 (2018)
10. Bródka, P., Filipowski, T., Kazienko, P.: An introduction to community detection in multi-layered social network. In: Lytras, M.D., Ruan, D., Tennyson, R.D., Ordonez De Pablos, P., García Peñalvo, F.J., Rusu, L. (eds.) WSKS 2011. CCIS, vol. 278, pp. 185–190. Springer, Heidelberg (2013). https://doi.org/10.1007/978-3-642-35879-1_23

11. Cardillo, A., et al.: Emergence of network features from multiplexity. Sci. Rep. **3**(1344), 1–6 (2013). https://doi.org/10.1038/srep01344. https://www.nature.com/articles/srep01344
12. Coleman, J., Katz, E., Menzel, H.: The diffusion of an innovation among physicians. Sociometry **20**(4), 253–270 (1957). http://www.jstor.org/stable/2785979
13. De Domenico, M., Lancichinetti, A., Arenas, A., Rosvall, M.: Identifying modular flows on multilayer networks reveals highly overlapping organization in interconnected systems. Phys. Rev. X **5**, 011027 (2015). https://doi.org/10.1103/PhysRevX.5.011027
14. Dickison, M.E., Magnani, M., Rossi, L.: Multilayer Social Networks. Cambridge University Press, Cambridge (2016). https://doi.org/10.1017/CBO9781139941907
15. Fortunato, S.: Community detection in graphs. Phys. Rep. **486**(3–5), 75–174 (2010)
16. Gottfried, J.: Americans' Social Media Use, vol. 31. Pew Research Center (2024)
17. Jankowski, J., Michalski, R., Bródka, P.: A multilayer network dataset of interaction and influence spreading in a virtual world. Sci. Data **4**(1), 170144 (2017). https://doi.org/10.1038/sdata.2017.144
18. Kamiński, B., Pankratz, B., Prałat, P., Théberge, F.: Modularity of the ABCD random graph model with community structure. J. Complex Netw. **10**(6), cnac050 (2022)
19. Kamiński, B., Prałat, P., Théberge, F.: Artificial benchmark for community detection (ABCD) - fast random graph model with community structure. Netw. Sci. 1–26 (2021)
20. Kamiński, B., Prałat, P., Théberge, F.: Mining Complex Networks. Chapman and Hall/CRC (2021). https://doi.org/10.1201/9781003218869
21. Kamiński, B., Prałat, P., Théberge, F.: Artificial benchmark for community detection with outliers (ABCD+o). Appl. Netw. Sci. **8**(1), 25 (2023)
22. Kamiński, B., Prałat, P., Théberge, F.: Hypergraph artificial benchmark for community detection (h–ABCD). J. Complex Netw. **11**(4), cnad028 (2023)
23. Kamiński, B., Olczak, T., Pankratz, B., Prałat, P., Théberge, F.: Properties and performance of the ABCDe random graph model with community structure. Big Data Res. **30**, 100348 (2022)
24. Kendall, M.G.: A new measure of rank correlation. Biometrika **30**(1–2), 81–93 (1938)
25. Kivelä, M., Arenas, A., Barthelemy, M., Gleeson, J.P., Moreno, Y., Porter, M.A.: Multilayer networks. J. Complex Netw. **2**(3), 203–271 (2014). https://doi.org/10.1093/comnet/cnu016
26. Lancichinetti, A., Fortunato, S.: Benchmarks for testing community detection algorithms on directed and weighted graphs with overlapping communities. Phys. Rev. E **80**(1), 016118 (2009)
27. Lancichinetti, A., Fortunato, S., Radicchi, F.: Benchmark graphs for testing community detection algorithms. Phys. Rev. E **78**(4), 046110 (2008)
28. Magnani, M., Hanteer, O., Interdonato, R., Rossi, L., Tagarelli, A.: Community detection in multiplex networks. ACM Comput. Surv. (CSUR) **54**(3), 1–35 (2021)
29. Magnani, M., Rossi, L., Vega, D.: Analysis of multiplex social networks with R. J. Stat. Softw. **98**, 1–30 (2021)
30. Omodei, E., De Domenico, M., Arenas, A.: Characterizing interactions in online social networks during exceptional events. Front. Phys. **3**, 59 (2015)
31. Orman, G.K., Labatut, V.: A comparison of community detection algorithms on artificial networks. In: Gama, J., Costa, V.S., Jorge, A.M., Brazdil, P.B. (eds.) DS 2009. LNCS (LNAI), vol. 5808, pp. 242–256. Springer, Heidelberg (2009). https://doi.org/10.1007/978-3-642-04747-3_20

32. Paradowski, M.B., Whitby, N., Czuba, M., Bródka, P.: Peer interaction dynamics and second language learning trajectories during study abroad: a longitudinal investigation using dynamic computational social network analysis. Lang. Learn. (2024)
33. Rossi, L., Magnani, M.: Towards effective visual analytics on multiplex and multilayer networks. Chaos Solitons Fractals **72**, 68–76 (2015). Multiplex Networks: Structure, Dynamics and Applications. https://doi.org/10.1016/j.chaos.2014.12.022. https://www.sciencedirect.com/science/article/pii/S0960077914002422
34. Snijders, T., Pattison, P.E., Robins, G.L., Handcock, M.S.: New specifications for exponential random graph models. Sociol. Methodol. **36**(1), 99–153 (2006). https://doi.org/10.1111/j.1467-9531.2006.00176.x
35. Tang, L., Wang, X., Liu, H.: Community detection via heterogeneous interaction analysis. Data Min. Knowl. Disc. **25**, 1–33 (2012)
36. Vinh, N.X., Epps, J., Bailey, J.: Information theoretic measures for clusterings comparison: is a correction for chance necessary? In: Proceedings of the 26th Annual International Conference on Machine Learning, pp. 1073–1080 (2009)
37. Wormald, N.C.: Generating random regular graphs. J. Algorithms **5**(2), 247–280 (1984)
38. Wormald, N.C., et al.: Models of random regular graphs. Lond. Math. Soc. Lect. Note Series 239–298 (1999)

Integrating Link Prediction and Isolation Forest for Backbone Extraction

Ali Yassin[1(✉)], Hocine Cherifi[2], Hamida Seba[3], and Olivier Togni[1]

[1] Laboratoire d'Informatique de Bourgogne - Université de Bourgogne, Dijon, France
ali.yassin@u-bourgogne.fr
[2] ICB UMR 6303 CNRS - Université de Bourgogne, Dijon, France
[3] Univ Lyon, UCBL, CNRS, INSA Lyon, LIRIS, UMR5205,
69622 Villeurbanne, France

Abstract. Backbone extraction simplifies complex networks by preserving essential connections while reducing complexity. Traditional methods rely on fixed thresholds, making them sensitive to parameter choices. Indeed, rare but structurally significant links are often discarded. To address this challenge, we propose a framework that integrates link prediction with anomaly detection. The method assigns scores to links using a similarity-based link prediction function. It then applies Isolation Forest to identify structurally significant links. Unlike conventional approaches, it retains high and low-scoring links, preserving strong connectivity patterns and rare but meaningful interactions. We validate the framework using two link prediction functions: Preferential Attachment and Local Path Index. Three experiments assess its effectiveness. First, we illustrate its ability to preserve central and peripheral links using Zachary's Karate Club network. Second, we compare it with traditional backbone extraction methods, including the Disparity Filter and the High Salience Skeleton, by evaluating edge and node retention, connectivity, reachability, transitivity, and clustering coefficient deviations. Results show that the Local Path Index backbone best maintains network connectivity and clustering, while the Preferential Attachment backbone provides a more condensed structure. The High Salience Skeleton causes excessive fragmentation, and the Disparity Filter removes too many links, leading to information loss. Third, we analyze how different methods alter edge weight and node degree distributions. The Local Path Index backbone best preserves the original degree distribution, while the Preferential Attachment backbone minimizes edge-weight distortions. The Disparity Filter heavily alters weight distributions, and the High Salience Skeleton significantly modifies degree distributions.The proposed framework balances connectivity and simplification, making it a flexible alternative to traditional backbone extraction methods.

Keywords: Backbone Extraction · Link Prediction · Anomaly Detection · Complex Networks · Isolation Forest

1 Introduction

Complex networks model diverse real-world systems, including social interactions, biological processes, and transportation networks. However, their inherent complexity often obscures critical relationships, making analysis challenging. Backbone extraction addresses this issue by simplifying network structures, retaining only the most significant connections while filtering noise. For instance, in biological networks, backbones identify key protein interactions essential for cellular function [16,21], while in finance, they reveal systemic risk pathways between financial institutions [8]. Similarly, social and infrastructural networks benefit from backbones that highlight vital connections shaping system dynamics.

Traditional backbone extraction methods fall into two categories: statistical and structural approaches [19,20]. Statistical methods rely on hypothesis testing or empirical distributions to assess the significance of edges or nodes [4,7,15]. They typically compute p-values and filter out statistically insignificant connections based on a predefined significance level. Structural methods, on the other hand, focus on network topology, leveraging shortest paths and connectivity patterns to extract meaningful substructures [5,6]. Some methods extract a single substructure, while others assign scores to nodes or edges, allowing threshold adjustments for greater flexibility.

While backbone extraction simplifies the network structure, link prediction offers another important perspective by estimating the likelihood of connections between nodes. Link prediction methods are typically classified into four categories [2]: similarity-based, dimensionality reduction-based, probabilistic and maximum likelihood based, and learning-based methods. Among these, similarity-based methods are the simplest and most fundamental. They rely on predefined similarity functions based on assumptions regarding edge formation to compute a similarity score to each pair of nodes. In this work, we focus on similarity-based approaches due to their simplicity and interpretability.

This paper introduces a novel backbone extraction framework that integrates link prediction with anomaly detection. The proposed approach uses the Isolation Forest algorithm [11] to identify key edges. Unlike threshold-dependent methods, it dynamically isolates anomalies without assuming data distributions, enhancing scalability and interpretability. Instead of treating anomalies as noise, the method preserves edges with extreme link prediction scores as structurally significant. This adaptive approach captures two complementary dimensions of network significance:

- **Central interactions:** High-scoring edges that represent the network's core structure.
- **Peripheral interactions:** Low-scoring edges that represent rare but contextually important ties.

We validate the framework through three key experiments: First, we illustrate the proposed framework and demonstrate how similarity functions expose interpretable backbone structures. This is exemplified with a toy example, Zachary's

Karate Club. Next, we conduct a comparative analysis against traditional backbone extraction methods, including the Disparity Filter [15] and High Salience Skeleton [6]. This analysis assesses edge, node retentions, reachability, number of components, transitivity, and average clustering coefficient deviations. Finally, we perform a distributional analysis of edge weights and node degrees. This analysis evaluates the conformity to the original network using Kolmogorov-Smirnov statistics.

This work presents a paradigm shift in backbone extraction, positioning anomalies as structurally vital rather than outliers. The proposed framework equips researchers and practitioners with a flexible tool. It allows them to distill complex networks into interpretable, functionally critical backbones. This approach enables deeper insights across domains.

The remainder of this paper is organized as follows: Sect. 2 reviews related work on backbone extraction, link prediction, and anomaly detection. Section 3 details the proposed methodology. Section 4 presents the datasets, evaluation metrics, and experimental setup. Section 5 illustrates the framework using a toy example. Section 6 compares the topological properties of the proposed backbone with traditional methods. Section 7 examines the weight and degree distributions of the extracted backbone. Finally, Sect. 8 concludes the study with future research directions.

2 Background

2.1 Backbone Extraction Methods

Backbone extraction simplifies complex networks by preserving core structural and functional features. The process reduces complexity for analysis and visualization while maintaining critical connections. Existing methods fall into two categories: statistical and structural approaches [19,20].

Statistical Methods use hypothesis testing or empirical weight distributions to identify significant edges. The Disparity Filter [15] is widely adopted. It assumes that a node distributes its total strength $s_u = \sum_{z \in \Gamma(u)} w_{(u,z)}$ uniformly across its edges. A one-sided right-tail test computes the probability α_{uv} of observing a weight W at least as extreme as $p_{(u,v)}$:

$$\alpha_{(u,v)} = 1 - (k_u - 1) \int_0^{p_{(u,v)}} (1-x)^{k_u - 2} \, dx$$

Here, k_u is the degree of u and $p_{(u,v)} = \frac{w_{(u,v)}}{s_u}$ is the normalized weight from u's perspective. Edges with $\alpha_{uv} \geq \alpha$ (the chosen significance level) are removed.

Structural Methods focus on the network topological properties to extract a backbone with desired structural properties. For instance, the High Salience Skeleton (HSS) [6] identifies edges critical to shortest paths across the network.

It relies on the definition of the edge salience $s_{(u,v)}$ to identify critical edges. For each node u, a shortest path tree T_u is constructed, where $T_{(u,v)}(u) = 1$ if edge (u, v) is part of the tree rooted at u, and 0 otherwise. Edge salience is then computed as:

$$s_{uv} = \frac{1}{|N|} \sum_{u \in N} T_{(u,v)}(u)$$

where $|N|$ is the total number of nodes. Thus, $s_{(u,v)}$ represents the fraction of shortest path trees containing edge (u, v). Edges with salience below a specified threshold are removed.

2.2 Link Prediction Methods

Link prediction is a fundamental problem that aims to estimate the likelihood of a connection between two nodes. Broadly, link prediction methods can be categorized into four main types [2,10]: similarity-based methods, probabilistic methods, dimensionality reduction-based methods, and learning-based methods. Among these methods, similarity-based methods, also known as heuristics methods, are flexible and easy to use and are the mainstay of link prediction algorithms. They rely on measures of node similarity to estimate the likelihood of a link. These methods can be further categorized into local, global, and quasi-local methods based on the scope of the information they use:

Local Methods leverage information from the immediate neighborhood of nodes to predict links. These methods are computationally efficient and straightforward to implement, making them suitable for large-scale networks. For instance, the Preferential Attachment assumes that nodes with higher degrees are more likely to form connections. It calculates the likelihood of a link based on the product of the degrees of the two nodes under consideration. The similarity score is computed as follows:

$$S(u, v) = k_{(u)} \cdot k_{(v)} = \sum_{z \in \Gamma(u)} w_{(u,z)}^{\alpha}$$

where $k_{(u)}$ represents the weighted degree of node u, $\Gamma(u)$ signifies the set of neighboring nodes to node u, and α is a parameter used to modify the impact of the edge weights.

Global Methods analyze the entire network structure to estimate edge probabilities, potentially improving accuracy. Typically, they exploit path structure, spectral properties, matrix decompositions, and random walks. These approaches built on well-established mathematical principles offer provable guarantees about their behavior and performance in link prediction. Yet, the substantial computational requirements challenge scaling up for larger networks.

Quasi-Local Methods provide a middle ground, balancing the efficiency of local methods and global approaches. By incorporating partial global information, these methods achieve enhanced predictive accuracy without incurring the computational overhead of fully global techniques. One of the popular methods is the Local Path Index. It assumes that nodes connected by a greater number of paths of varying lengths have a higher likelihood of forming new connections. The similarity score is computed as:

$$S(u,v) = |paths^2_{(x,y)}| + \epsilon \cdot |paths^3_{(x,y)}|$$

where, ϵ is a free parameter. When ϵ is zero, the metric simplifies to the common neighbor method. And $|paths^3_{(x,y)}|$ represents the total number of different routes of length 3 between them.

Local and semi-local methods achieve an optimal balance between scalability, memory efficiency, adaptability, and predictive performance. They are particularly well-suited for large, sparse networks, while global approaches are computationally prohibitive. Given these efficiency properties, this study focuses on local and quasi-local methods.

2.3 Anomaly Detection Methods

Anomaly detection is a fundamental process that aims to identify unusual patterns in data. It has applications in fraud detection, network intrusion detection, manufacturing quality control, and healthcare analytics. These applications benefit from prompt detection of unusual patterns that may indicate threats, errors, or emerging trends. Approaches for anomaly detection often fall into distance-based, density-based, or model-based methods [3,12]. Among these methods, the model-based methods are effective at capturing normal behavior. They build models that learn typical patterns and then detect data points that deviate from these learned patterns. This data-driven approach often reduces the need for manual thresholding.

Isolation Forest Algorithm [11] is a model-based anomaly detection method. The model constructs an ensemble of random trees. Each tree is grown by splitting data points into random features and random split values. Points that require fewer splits to become isolated are more anomalous. The anomaly score $s(x,n)$ is calculated as follows:

$$s(x,n) = 2^{-\frac{E[h(x)]}{c(n)}}$$

where $E[h(x)]$ is the average path length of a point x across all trees. n is the number of samples. Let $c(n)$ be a normalizing constant, often approximated by $2H(n-1) - \frac{2(n-1)}{n}$, where H is the harmonic number. A score near 1 indicates a higher chance of being an anomaly. A score much lower than 1 suggests normal behavior. The method does not depend on specific distribution assumptions and scales well to large, high-dimensional datasets.

3 Backbone Extraction Framework

The proposed framework integrates link prediction with anomaly detection. It uses the isolation forest method to identify the edges of the backbone. It does this by reinterpreting anomalies as structurally important, rather than merely outlier edges. Figure 1 shows the process.

The process begins by iteratively evaluating each observed edge e. For every edge, it is temporarily removed from the network. Subsequently, we evaluate the probability of the edge's existence using a similarity-based metric. After scoring, the edge is reinserted to preserve the original network structure.

The raw scores undergo logarithmic transformation to reduce skewness and are normalized to a standardized range (e.g., [0, 1]) for consistency. Next, we feed these values into an Isolation Forest model. The model isolates edges with unusual patterns. Edges that stand out as "anomalies" are preserved as the backbone edges, indicating their unique role in the network.

The proposed method is robust and scalable, as it does not rely on predefined thresholds or assumptions about the network's structure. Instead, it adaptively identifies backbone edges based on the distribution of similarity scores derived from the chosen link prediction function. This flexibility allows the framework to enhance interpretability, scalability, and applicability across domains, making it suitable for both small-scale networks and large, complex systems.

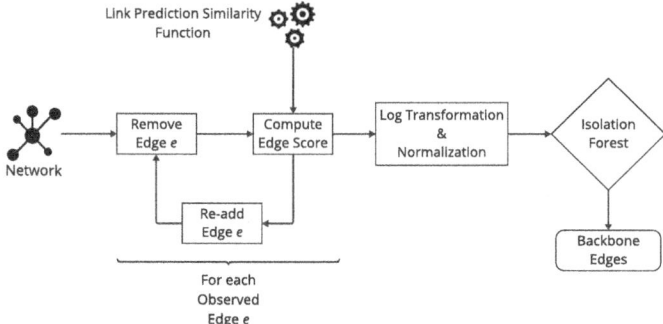

Fig. 1. A diagram illustrating the backbone extraction process within the proposed framework. Each observed edge is temporarily removed, and its existence probability is estimated using a similarity-based metric. The edge is then reinserted to maintain the network structure. The computed scores undergo log transformation and normalization. Finally, Isolation Forest analyzes the scores, identifying anomalous edges, which are preserved as backbone edges.

4 Data, Materials, and Methods

4.1 Dataset

The experiment involves three real-world networks: infrastructural, social, and character. Table 1 reports their basic topological features.

- **Global Airports** [1]: Nodes represent world airports, edges denote direct flights between them, and edge weights indicate the number of flights operated by various airlines from May 17 to May 22, 2018.
- **Les Misérables** [9]: Nodes depict characters from Victor Hugo's novel "Les Misérables," connected if they appear together in the same chapter. Edge weights indicate how many times such co-appearances occur in the novel.
- **Zachary's Karate Club** [17]: Nodes represent individuals from Zachary's Karate club. An edge exists between two nodes if the corresponding individuals are seen together outside of normal club activities. Edge weights represent the number of such occurrences.

Table 1. The network topological features: $|N|$, $|E|$, $<k>$, and ρ are the number of nodes, edges, average degree, and density.

Network	$|N|$	$|E|$	$<k>$	ρ
Global Airports	2,734	16,665	12.19	0.004
Les Misérables	77	254	6.59	0.087
Zachary's Karate Club	34	78	4.588	0.139

4.2 Evaluation Metrics

In this subsection, we introduce the evaluation metrics that are used to assess the performance of our proposed methods. These metrics are usually used for assessing backbone extraction methods.

Relative Difference Deviation. The Relative Difference Deviation measures the relative difference ΔP between a property P in the backbone network and the original network. It provides a normalized way to quantify changes in network properties after applying backbone extraction. It is defined as:

$$\Delta P = \frac{|P_{\text{backbone}} - P_{\text{original}}|}{P_{\text{original}}}$$

where P_{backbone} is the value of the property P in the backbone, and P_{original} is the value of the property P in the original network.

Reachability. The Reachability [13] quantifies the connectivity between all pair of nodes in a network. It is defined as the fraction of node pairs that can communicate with each other. This reads:

$$R = \frac{1}{n(n-1)} \sum_{i \neq j \in G} R_{ij}.$$

with n is the number of nodes and $R_{ij} = 1$ if path exists between node i and j and $R_{ij} = 0$ otherwise. The Reachability values are in the $[0, 1]$ range. If any pair of nodes can communicate in a network, the reachability R becomes 1. If $R = 0$ it means all nodes are isolated from each other.

Average Clustering Coefficient. The average clustering coefficient C quantifies the tendency of nodes in a network to cluster together [14]. It is calculated as the mean of the local clustering coefficients of all nodes. The local clustering coefficient of a node is defined as the fraction of triangles it forms with its neighbors relative to the total number of possible triangles. This reads:

$$C = \frac{1}{n} \sum_{i=1}^{n} c_i$$

where n is the total number of nodes in the graph, and $c_i = \frac{2 \cdot T(i)}{k_i \cdot (k_i - 1)}$ is the local clustering coefficient of node i, $T(i)$ is the number of triangles involving node i and k_i is the degree of node i.

Transitivity. The transitivity [18] of a network reflects how likely neighboring nodes are connected. Mathematically, it's computed as the ratio of the number of triangles (Δ) to the number of connected triples of nodes (τ):

$$T = \frac{3 \times \Delta}{\tau}$$

4.3 Experimental Setup

This subsection describes the experiments conducted to evaluate the proposed framework and compare it with existing methods.

The first experiment illustrates the proposed framework in a toy example. We extract the backbone of the popular Karate Club network using the preferential attachment and local path similarity function. Then we interpret the extracted backbone. Finally, we examine the distribution of normalized scores and the backbone edges.

The second experiment compares the proposed framework with traditional backbone extraction methods: the Disparity Filter (statistical) and the High Salience Skeleton (structural) methods. We extract two backbones for our framework using the Preferential Attachment and the Local Path Index similarity functions. We set a typical significance level $\alpha = 0.05$ for the Disparity Filter. We set the High Salience Skeleton threshold to 0.8, as recommended by its authors. After extracting all backbones, we compare topological properties including: the fraction of edges and nodes retained, the number of components, reachability, and deviations in transitivity and average clustering coefficient relative to the original network.

Finally, we examine the weight and degree distributions for the extracted backbones. We plot the cumulative distributions for each backbone and compare them to the original network. We then compute the Kolmogorov-Smirnov (KS) statistic to measure the distance between each backbone's distribution and the original distribution.

5 Toy Example Illustration

To demonstrate the proposed framework, we apply it to Zachary's Karate Club network. The network is a well-studied social network representing interactions between karate club members outside the club. During the study, a conflict arose between the administrator "John A" (node 33) and the instructor "Mr. Hi" (node 0), ultimately leading to the club splitting into two factions. We extract two backbones using two different similarity functions: Preferential Attachment and Local Path Index.

Figure 2 illustrates the extracted backbones. The upper left and lower left panels show the original network. The edge colors represent the Preferential Attachment and Local Path Index similarity scores, respectively.. The upper right and lower right panels display the corresponding extracted backbones.

The Preferential Attachment backbone retains 17 nodes (50% of the original network) and 12 edges (15.4% of total edges). In contrast, the Local Path backbone maintains a slightly larger subset of the network. It preserves 20 nodes (58.8% of the original network) and 20 edges (25.6% of total edges),

To understand the extracted backbones, it is essential to examine the edge scores, as they are derived from different underlying assumptions. In both backbones, we observe that edges with extreme normalized scores-both high-scoring (yellow) and low-scoring (blue)-are retained. However, the interpretation of these backbones differs based on their respective similarity functions.

In the Preferential Attachment Backbone High-score edges represent interactions between high-degree (popular) nodes, which frequently connect with others. For example, edges between the instructor (node 0) and nodes 1, 2, and 31. In contrast, low-score edges represent interactions between low-degree (non-social) nodes that rarely interact with others. For instance, node 11 only interacts with the instructor (node 0) throughout the study, making it an isolated link.

In Local Path similarity assumes that nodes connected by multiple short paths are more likely to form direct links. Thus, high-score edges are those strongly supported by redundant local paths (e.g., multiple routes of length two or three). For example, interactions between the instructor (node 0) and nodes 1, 2, and 13, which form a fully connected clique in the original network. In contrast, low-score edges lack strong support from redundant local paths,

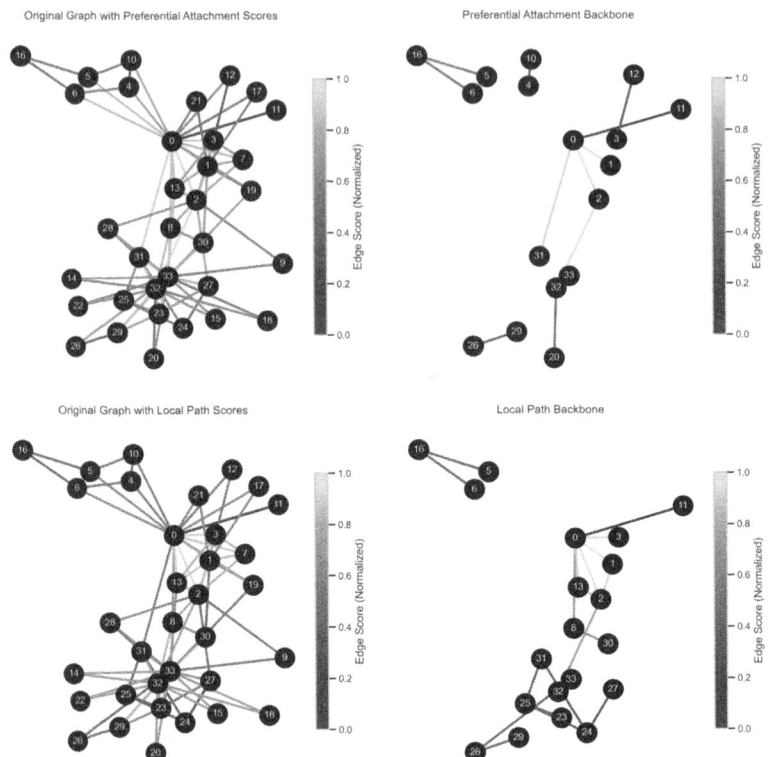

Fig. 2. The Zachary Karate Club original network (left) with edge colors reflecting the normalized similarity scores. The extracted Preferential Attachment (PA) backbone (top right) and Local Path (LP) backbone (bottom right).

often appearing on the periphery of the network. These are typically interactions between sparsely connected nodes, such as node 0 with nodes 11, 27, and 24.

Figure 3 further validates these findings by displaying the distribution of normalized scores, with backbone edges highlighted in red. In both cases, the Isolation Forest identifies edges at both extremes of the score distribution as backbone edges. However, in the Local Path backbone (right panel), the algorithm also selects some mid-range score edges as part of the backbone. This behavior can be explained by the low frequency of these mid-range values in the network. Since Isolation Forest isolates data points based on how easily they can be separated, sparsely occurring edges require fewer splits to become isolated. This makes them more likely to be classified as structurally significant, even if their similarity scores are moderate.

By selecting edges from both ends of the similarity spectrum, the framework captures both dominant structures and hidden interactions. High-score edges maintain the core hierarchy of the network, reinforcing strong, well-established connections. Low-score edges highlight less obvious but functionally relevant interactions that might otherwise be overlooked. Additionally, the framework may isolate rare edges with mid-range values that deviate from typical patterns. While their exact structural role is uncertain, their rarity suggests they differ from more frequent connections in a meaningful way. This aligns with real-world network patterns, where a dominant structure coexists with unexpected or peripheral interactions, which are crucial for network resilience, adaptability, and emergent dynamics.

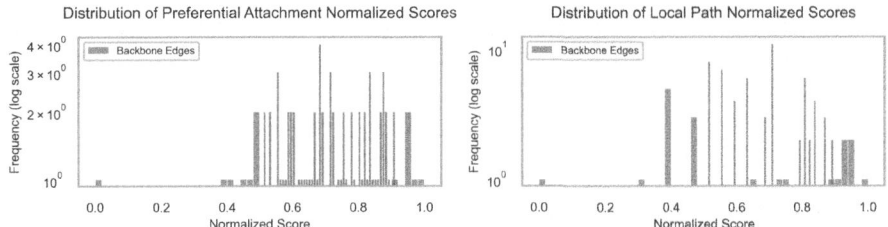

Fig. 3. The distribution of normalized scores computed using the Preferential Attachment similarity function in the Zachary's Karate Club network. The backbone edges highlighted in red. (Color figure online)

6 Comparing Backbones Topological Properties

Table 2 reports the topological properties of the extracted Preferential Attachment (PA), Local Path Index (LP), High Salience Skeleton (HSS), and Disparity Filter (DF) backbones in two networks: the Global Airports and Les Misérables.

Table 2. The topological properties of the extracted Preferential Attachment (PA), Local Path Index (LP), High Salience Skeleton (HSS), and Disparity Filter (DF) backbones in two networks: the Global Airports and Les Misérables. Bold values represent the best performance.

Property	Global Airports				Les Misérables			
	PA	LP	HSS	DF	PA	LP	HSS	DF
Edge Percentage	0.26	0.30	0.15	0.10	0.196	0.20	0.29	0.04
Node Percentage	0.84	0.91	**0.95**	0.29	0.71	0.70	**0.97**	0.13
Reachability	0.63	**0.95**	0.02	0.83	0.17	0.93	**0.95**	0.64
Number Components	79	24	159	**19**	11	1	1	1
Transitivity Deviation	0.47	**0.06**	-1	-0.82	-0.67	-0.73	-1	**-0.57**
Average Clustering Deviation	-0.61	**-0.53**	-1	-0.89	-0.79	**-0.63**	-1	-0.77

The primary goal of backbone extraction methods is to reduce the size of a network. Here, the Disparity filter consistently retains the lowest edge fraction in both networks (10% and 4%). In contrast, the Local Path (PA) method preserves the highest fraction in Global Airports (30%). For the Les Misérables, the High Salience Skeleton (HSS)retains the most edge fraction (29%).

The preservation of nodes is not always an ideal characteristic of a filtering technique. However, maintaining network nodes throughout filtering poses a considerable challenge for many techniques. Here, the High Salience Skeleton (HSS) outperforms all methods, retaining 95% of nodes in Global Airports and 97% in Les Misérables. In contrast, the Disparity filter (DF) performs worst, keeping only 29% and 13% in the Global Airports and Les Misérables, respectively. The Preferential Attachment (PA) and Local Path (LP) methods lie between these two extremes. They preserve a decent node fraction that ranges between 70% and 90%.

Filtering the network can potentially disrupt the connections within the network, leading to a backbone composed of multiple components. Therefore, it is crucial to evaluate backbone extraction techniques by comparing the reachability of nodes in the extracted backbones. There is no single method that maintains high reachability across both networks. The Local Path (LP)) achieves near-perfect reachability (0.95) in Global Airports, while the High Salience Skeleton (HSS) extracts collapses to 0.02. In Les Misérables, the High Salience Skeleton (HSS) and Local Path (LP) extract backbones with a high reachability (0.95 and 0.93). The Disparity Filter (DF) shows moderate performance (0.83 in Airports, 0.64 in Les Misérables).

Low reachability values indicate network disconnection but do not clarify how the network is fragmented. Hence, we examine the number of components in the extracted backbones to better understand their connectivity. The Disparity Filter (DF) extracts a backbone with the fewest number of components in Global Airports (19). In contrast, the High Salience Skeleton (HSS) fragments the network into 159 components. In Les Misérables, the Local Path (LP), High Salience Skeleton (HSS), and Disparity Filter (DF) ideally maintain a single component. However, the Preferential Attachment (PA) splits the network into 11 components.

Transitivity in networks refers to the tendency for nodes with a mutual connection to also be connected. Filtering edges may disrupt these cycles, fragmenting local structures. The High Salience Skeleton (HSS) diminishes transitivity most severely (deviation $= -1.00$ in both networks). The Local Path (LP) backbone almost preserves transitivity in the Global Airports (-0.06). The Disparity Filter (DF) performs best in Les Misérables (-0.57).

The average clustering coefficient quantifies local cohesion by measuring how tightly nodes cluster with their neighbors. Deviations in this metric after backbone extraction reflect the preservation (or loss) of community-like structures. The results mirror the transitivity trends. The High Salience Skeleton (HSS) exhibits the largest negative deviations (-1.00 in both networks), eroding local clustering by prioritizing global connectivity. The Local Path (LP) backbone minimizes disruption, with deviations of −0.53 and −0.63, outperforming other methods. This aligns with its focus on local connectivity patterns.

7 Comparing Backbones Distributions

This section evaluates the proposed framework against two traditional backbone extraction methods-the Disparity Filter (DF) (statistical) and High Salience Skeleton (HSS) (structural)-by analyzing their ability to preserve the original network 's weight and degree distributions. For the proposed framework, the backbones were extracted using preference attachment (PA) and local path index (LP) similarity functions.

Node degrees reflect connectivity patterns critical to network function. Effective filtering should retain diverse degrees to preserve the original distribution. Figure 4 illustrates the cumulative distributions of node degrees for all backbones are compared to the original network, with the Kolmogorov-Smirnov (KS) statistic quantifying deviations. All methods alter the original degree distribution, but to varying extents. The High Salience Skeleton (HSS) backbone exhibits the largest deviation in the Global Airports and Les Misérables networks. In contrast, the Local Path (LP) backbone of the proposed framework achieves the closest alignment. It has the lowest KS statistic 0.194 in Global Airports and 0.421 in Les Misérables.

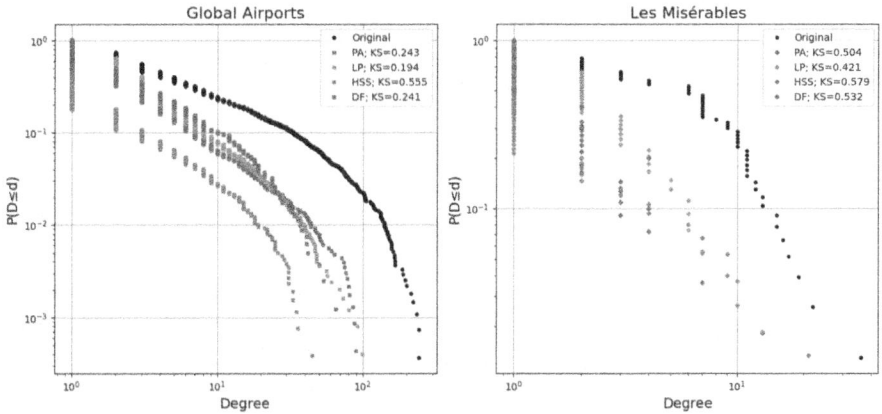

Fig. 4. The increasing cumulative distributions of edge degrees for the original network and extracted backbones with the Kolmogorov-Smirnov (KS) statistic in the legend. The backbones are: Disparity (DF), High Salience Skeleton (HSS), and the proposed Local Path (LP) and Preferential Attachment (PA).

Edge weights encode the strength of the relations between the entities. Effective filtering should retain different weight scales to preserve the original distribution. Figure 5 illustrates the cumulative distributions of edge weights for all backbones are compared to the original network, with the Kolmogorov-Smirnov (KS) statistic quantifying deviations. The Disparity (DF) backbone deviates significantly, failing to retain weight scales in both networks. In contrast, the High Salience Skeleton (HSS) backbone performs best in the Global Airports network (KS = 0.148). Followed by the proposed Local Path (LP) backbone with a moderate KS values (0.203). Similarly, the proposed Preferential Attachment (PA) backbone excels in Les Misérables (KS = 0.101).

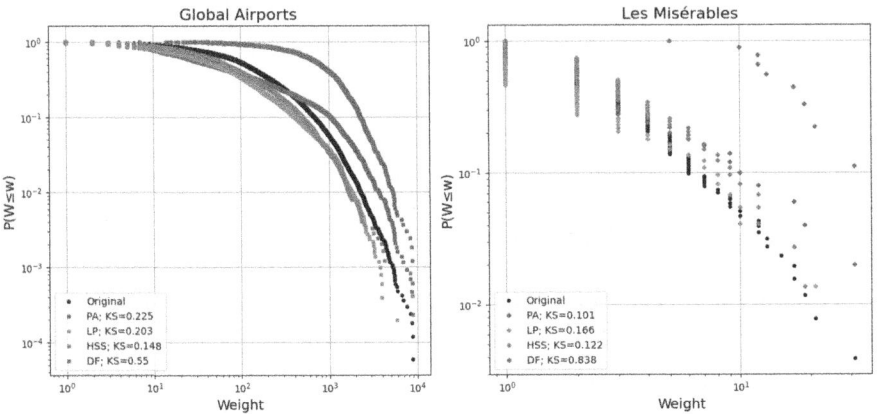

Fig. 5. The cumulative distributions of edge weights for the original network and extracted backbones with the Kolmogorov-Smirnov (KS) statistic in the legend. The backbones are: Disparity (DF), High Salience Skeleton (HSS), and the proposed Local Path (LP) and Preferential Attachment (PA).

8 Conclusion

This study introduces a novel framework for backbone extraction in complex networks by reimagining anomaly detection through the lens of structural significance. The approach combines link prediction similarity functions with the Isolation Forest algorithm to detect anomalies as backbone edges. These edges are characterized either by being vitally central (indicated by high similarity scores) or distinctly peripheral (indicated by low similarity scores). This dual focus captures both the dominant hierarchical structure and the nuanced, often overlooked interactions that shape real-world networks.

The choice of similarity function shapes the interpretation of the extracted backbone. In the Karate Club toy example, the Preferential Attachment highlights the social hierarchy. In contrast, the Local Path Index emphasizes redundant short paths and localized connectivity.

Comparative analysis against Disparity Filter (DF) and High Salience Skeleton (HSS) shows no universally superior method. The Local Path (LP) backbone balances global reachability, local clustering, and degree distribution. While Preferential Attachment (PA) and High Salience Skeleton (HSS) aligns more closely with the original weight distribution. The High Salience Skeleton (HSS) excels at retaining the original nodes but deviates most from the degree distribution.

The scalable and threshold-free framework design allows dynamic backbone extraction across diverse networks. Its flexibility in the selection of similarity functions enables tailored interpretations, making it a versatile tool for network analysis. Future work could explore additional global or learning-based link prediction methods to further enhance adaptability.

Acknowledgment. This material is based upon work supported by the Agence Nationale de Recherche under grant ANR-20-CE23-0002.

References

1. Alves, L., Aleta, A., Rodrigues, F., Moreno, Y., Amaral, L.: Centrality anomalies in complex networks as a result of model over-simplification. New J. Phys. (2020)
2. Arrar, D., Kamel, N., Lakhfif, A.: A comprehensive survey of link prediction methods. J. Supercomput. **80**(3), 3902–3942 (2024)
3. Chandola, V., Banerjee, A., Kumar, V.: Anomaly detection: a survey. ACM Comput. Surv. (CSUR) **41**(3), 1–58 (2009)
4. Foti, N.J., Hughes, J.M., Rockmore, D.N.: Nonparametric sparsification of complex multiscale networks. PLoS ONE **6**, e16431 (2011)
5. Ghalmane, Z., Cherifi, C., Cherifi, H., El Hassouni, M.: Extracting modular-based backbones in weighted networks. Inf. Sci. **576**, 454–474 (2021)
6. Grady, D., Thiemann, C., Brockmann, D.: Robust classification of salient links in complex networks. Nat. Commun. **3**, 864 (2012)
7. Hmaida, S., Cherifi, H., El Hassouni, M.: A multilevel backbone extraction framework. Appl. Netw. Sci. **9**(1), 41 (2024)
8. Keller-Ressel, M., Nargang, S.: The hyperbolic geometry of financial networks. Sci. Rep. **11**(1), 4732 (2021)
9. Knuth, D.: The Stanford GraphBase. A platform for combinatorial computing (1993)
10. Kumar, A., Singh, S.S., Singh, K., Biswas, B.: Link prediction techniques, applications, and performance: a survey. Phys. A **553**, 124289 (2020)
11. Liu, F.T., Ting, K.M., Zhou, Z.H.: Isolation forest. In: 2008 Eighth IEEE International Conference on Data Mining, pp. 413–422. IEEE (2008)
12. Pang, G., Shen, C., Cao, L., Van Den Hengel, A.: Deep learning for anomaly detection: a review. ACM Comput. Surv. (CSUR) **54**(2), 1–38 (2021)
13. Sato, Y., Ata, S., Oka, I.: A strategic approach for re-organization of internet topology for improving both efficiency and attack tolerance (2008)
14. Schank, T., Wagner, D.: Approximating clustering-coefficient and transitivity, Interner Bericht. Fakultät für Informatik, Universität Karlsruhe, vol. 2004. Universität Karlsruhe (TH) (2004)
15. Serrano, M.A., Boguna, M., Vespignani, A.: Extracting the multiscale backbone of complex weighted networks. Proc. Nat. Acad. Sci. **106**, 6483–6488 (4 2009). https://doi.org/10.1073/pnas.0808904106

16. Serrano, M., Boguñá, M., Sagucafés, F.: Uncovering the hidden geometry behind metabolic networks. Mol. BioSyst. **8**(3), 843–850 (2012)
17. Shore, K.A.: Complex networks: principles, methods and applications. Contemp. Phys. (2018). https://doi.org/10.1080/00107514.2018.1450296
18. Wasserman, S., Faust, K.: Social network analysis: methods and applications (1994)
19. Yassin, A., Cherifi, H., Seba, H., Togni, O.: Backbone extraction through statistical edge filtering: a comparative study. PLoS ONE **20**(1), e0316141 (2025)
20. Yassin, A., Haidar, A., Cherifi, H., Seba, H., Togni, O.: An evaluation tool for backbone extraction techniques in weighted complex networks. Sci. Rep. (2023)
21. Zádor, Z., Zhu, Z., Smith, M., Gorgoni, S.: A weighted and normalized Gould-Fernandez brokerage measure. PLoS ONE **17**(9), e0274475 (2022)

Author Index

A
Ardickas, Daumilas 65

B
Barrett, Jordan 81, 125
Bloznelis, Mindaugas 65
Boldi, Paolo 1
Bonato, Anthony 17
Bondar, Maksym 141
Bródka, Piotr 172
Bryskina, Veronika 141

C
Cherifi, Hocine 189
Czuba, Michał 172

D
DeWolfe, Ryan 81, 125

F
Furia, Flavio 1

H
Henning, Florian 96

K
Kamiński, Bogumił 81, 125, 172
Kraiński, Łukasz 172
Krilavičius, Tomas 141

L
Liang, Qiu 109
Litvak, Nelly 96, 109

M
Mandravickaitė, Justina 141
Meyer, François G. 157
Mörters, Peter 45

P
Prałat, Paweł 81, 125, 172
Prezioso, Chiara 1

R
Rybarczyk, Katarzyna 30

S
Schätze, Lucas 45
Seba, Hamida 189
Smith, Aaron 81, 125
Songailaitė, Milita 141

T
Théberge, François 81, 125, 172
Togni, Olivier 189

V
Vaicekauskas, Rimantas 65
van der Hofstad, Remco 96, 109

W
Walaa, Mariam 17

Y
Yassin, Ali 189

The manufacturer's authorised representative in the EU is Springer Nature Customer Service Centre GmbH, Europaplatz 3, 69115 Heidelberg, Germany. If you have any concerns regarding our products, please contact ProductSafety@springernature.com

Printed and bound by CPI Group (UK) Ltd, Croydon, CR0 4YY
25/03/2026
02078191-0011